Geological Ramblings in Yosemite

Geological Ramblings in Yosemite

N. King Huber

foreword by Jim Snyder

YOSEMITE ASSOCIATION,
Yosemite National Park, California

HEYDAY BOOKS,
Berkeley, California

Copyright © 2007 by the Yosemite Association.

All rights reserved. No part of this book may be reproduced or transmitted in any form or by any means, electronic or mechanical, including photocopying and recording, or by any information storage or retrieval system, without permission in writing from Heyday Books.

Library of Congress Cataloging-in-Publication Data

Huber, N. King (Norman King), 1926-
  Geological ramblings in Yosemite / N. King Huber; foreword by Jim Snyder.
    p. cm.
  ISBN 978-1-59714-072-0 (pbk. : alk. paper)
 1. Geology--California--Yosemite National Park. 2. Natural history--California--Yosemite National Park. I. Title.
  QE90.Y6H78 2007
  557.94'47--dc22
                                         2007009805

Cover Photo by Keith S. Walklet
Back Cover Photo of Mount Hoffmann in 1867 by W. Harris,
   courtesy of the Bancroft Library, University of California, Berkeley
Author Photo by Nakata/Meyer: Sight & Sound Productions © 1990
Cover and Interior Design by Lisa Buckley
Printing and Binding: CDS Publications, Medford, OR

Orders, inquiries, and correspondence should be addressed to:
    Heyday Books
    P.O. Box 9145, Berkeley, CA 94709
    (510) 549-3564, Fax (510) 549-1889
    www.heydaybooks.com

Printed in the United States of America

10 9 8 7 6 5 4 3 2 1

For my parents,
Norman and Marie, dedicated Mountaineers, who from the
very beginning instilled in me a love for the outdoors.
They would have been delighted that this eventually would
lead from their own western mountain, Mount Rainier,
to my Sierra Nevada and Yosemite.

Contents

Foreword by Jim Snyder   ix

Introduction
 Mountain Reminiscences   3

"Incomparable Valley"
 The Geologic Story of Yosemite Valley   15

Exploring Yosemite Geology
 Interpreting Yosemite Geology, A Historical Perspective   33
 A Glacial Footnote   49

Tracking the Fire
 Evolution of the Tuolumne River   57
 James Mason Hutchings and the Devils Postpile   65

Tracking the Ice
 From V to U—Glaciation and Valley Sculpture   73
 A Tale of Two Valleys   79
 How Deep Is the Valley?   86

After the Ice
 Yosemite Falls—A New Perspective   95
 A History of the El Capitan Moraine   103
 Exotic Boulders at Tioga Pass   111
 The Slide   115

Foreword

IN THE DAYS WHEN UNIVERSITIES HAD BREADTH REQUIREMENTS TO ENCOURage a well-rounded education, I had to take several courses in my weak suit, the sciences. Unlike King Huber, who began his college career with a major in physics, I was sure I would fail any such subject. So, with graduation coming soon, I finished my science requirement by taking a basic course in geology. Professor Charles G. Higgins taught the course that spring in 1965. It was an interesting course, and I survived it. What I remember most was Higgins' departure from the standard textbook fare near the end of the course to tell us about a controversial theory in geology that he felt in time would revamp the science: plate tectonics. It was a week of instruction I have never forgotten.

I wanted to teach college history, but every teaching job I landed fell through after a semester or two. Returning to Yosemite instead to work trails put me face to face with basic geology every day whether dealing with rockfall, trail erosion problems, or explosives. It was quite clear to me that I had little clue about any of these things, and, like most people working on trails, I accumulated experience mostly devoid of understanding. This sounds like a recipe for disaster, but as long as we dealt with smaller issues, we were all right. When bigger events occurred, however, like the 1980 rockfalls, something had to change. Yosemite had thought to solve the severe Yosemite Falls Trail damage that year by building a substitute trail up Indian Canyon. The plan did not take geology into account, and I was fearful of taking a crew into a canyon with frequent slides. Yosemite sent me to the U.S.

Geological Survey in Menlo Park that winter for help, and I began my next geology course. This time I was the lone student in a roomful of professors, all of whom had been working on the rockfall problem to prepare for my arrival. That was an unforgettable course, too. Among other things, it was my introduction to a number of people, including King Huber, who had been immersed for many years in the geology of the Sierra Nevada and Yosemite.

Not only were these geologists available for advice, they also frequented the park, even on vacations, looking at its geology. They did not mind that I had so little formal knowledge of their science: they were still interested in what I observed in the field. "I will tell you what I saw, and then you tell me what I really saw," became a joke between King and other geologists and me. They added immeasurably to the understanding so lacking in my day-to-day experience "on the rock pile," as we often put it. King and others shared their publications with us, introducing us to scientific articles and geologic maps and relating them to our trail work. We, of course, were obliged to send occasional rock samples back. One of those I sent to King in the early 1980s was from The Slide. He wanted a sample with lichen on it for possible dating, "and make sure it's a big enough sample to be useful," he said with a grin. King and Clyde Wahrhaftig both visited trail camps in the backcountry and came out on the trail to help us understand better what we were dealing with. They liked it when we broke rocks up so they could see inside, but they might also ask us not to destroy this boulder or that one because it had something to say about the geology of the place. "Leave it as a geological exhibit, if you can." Or they'd complain with a laugh that packers using rocks to balance their loads often made geological study difficult by bringing exotic rocks into places they weren't supposed to be.

Maybe it was the scale on which they worked that was so intriguing. While we focused on smaller things on the trail, King could look at a pebble one minute and the batholith the next. We looked at joints in the immediate areas of our work until King brought out aerial photos and a viewer to show us the massive regional joints that influenced so much of the topography around us (FIG. 1). I remember standing on a bluff above Pleasant Valley with King looking across to the V-shaped channel remnant of the old Tuolumne River filled with lava. At our feet was broken glacial polish. King said, "Here at our feet are ten thousand years of erosion in granite." Then he swept his arm toward the Grand Canyon of the Tuolumne and said, "And there are ten million years of erosion in granite." It took my breath away.

In all this, King had a sense of the history of his science. He frequently tracked down the earliest geological reports on a research area or type of rock. He loves to point out that Henry Ward Turner had taken the first photograph of that ancient Tuolumne River channel above Pleasant Valley and that Turner also had originally named some of the rock units there, had a name for Smedberg Lake before Smedberg came along, and was the namesake for Turner Lake in the Clark Range. King took a special interest in Josiah Whitney, who had been a predecessor in studying not only the geology of Yosemite and the Sierra but also of Isle Royale in Lake Superior, which King also mapped. There was much to learn from those early observers, over and beyond what theories they might have propounded. Maybe that's why King's favorite historic photograph shows Whitney Survey topographer Charles Hoffmann with his tripod and transit on top of Mt. Hoffmann; an original print hangs on a wall in his home.

While King has always enjoyed his geological work, he has also believed that the translation of scientific study to the general public is essential. Such explanation tested fieldwork and theory, sometimes sending him back to the drawing board or into a new research project, knowing all the time that there are no truly final conclusions and that new explanations will be developed over time, just like plate tectonics. In Yosemite, King has always had something like the Grouse Creek stream capture and Sherwin moraine to examine after a walk in the field with Park Service interpreters, resource people, trail crews, Yosemite Institute instructors, Yosemite Association members, or others. Like his good friend Clyde Wahrhaftig, King tried to include everyone interested: "Yesterday was a research day; today's a teaching day. Let's go for a walk."

That's what this book is, "a teaching day," sharing some of King's geological ramblings—with a bow to Joseph LeConte's *Journal of Ramblings* through the High Sierra of California (1875). King graciously consented to include a short geological autobiography as an introduction. In "Incomparable Valley," he sums up Yosemite Valley geology, laying the groundwork for the rest of the book. "Exploring Yosemite Geology" shares his thoughts about the history of geological study and mapping in Yosemite from Whitney's California Geological Survey to the USGS. "Tracking the Fire"

FIGURE 1
In Neall Meadow, King checks his field observations with aerial photographs of glacial geology in Rodgers Canyon, 1989. *Dwight Barnes photo.*

examines some volcanic problems from Yosemite, including the ancient Tuolumne River, while "Tracking the Ice" includes several studies of glaciation. "After the Ice" is about Yosemite geology once the glaciers receded. Here you have the benefit of a lifetime of geologic study to complement his earlier book, *The Geologic Story of Yosemite National Park*.

King's interest in Yosemite, as well as that of the USGS, have been extraordinarily fortunate for the rest of us. King and I share the hope that this book will make its readers want to take some geological ramblings of their own.

Jim Snyder
August 18, 2006

POSTSCRIPT

Geological Ramblings was not supposed to be King's last book. Nor did he think he had taken his last trip to the mountains he loved. After he was diagnosed with cancer in summer 2006, he was weakened by the long treatments but did not stop working. In December King wrote, "My being house-bound for so long allowed me to work on completing my latest book . . . This will be a fitting cap to my Yosemite career." He worked on a laptop provided by his sons in front of the fireplace, which became a reminder of many high country campfires. King passed away February 24, 2007. *Geological Ramblings* is his last gift to those who love Yosemite. He signed it for you this way:

Enjoy Yosemite with me -- as I know you will!

King Huber

Introduction

Mountain Reminiscences

In June of 1955 my wife and I drove up Lee Vining Canyon over Tioga Pass and into Yosemite. This was my introduction to the Sierra Nevada, and having grown up in the relatively flat terrain of northern Minnesota, a mountain of any kind was new to me. Little did I know what this magnificent mountain range would hold for me in the years to come.

This excursion resulted from my reassignment by the U.S. Geological Survey (USGS) from Michigan's Upper Peninsula, where I had been working on iron deposits, to the Survey's newly established regional office in Menlo Park, California. I had been a geologist with the Survey for only a year and looked forward to this new venue, even though I did not know exactly what my assignment would entail.

Mountains actually were in my heritage, although it took a long time for them to come to the fore. My parents met as members of the Seattle Mountaineers, and, one year in the early 1920s when they managed a remote back country summer camp on the north slopes of Mount Rainier, my father guided climbers to its summit (FIG. 1). At the end of the season they circled Mount Rainier with two saddle horses and a pack horse, and then continued 2½ days further on horseback to Seattle where they lived at that time.

My father's work as engineer for a company that built hoisting equipment for use in the logging industry eventually took them to Duluth, Minnesota, my birthplace. We lived on the shore of Lake Superior where one of my favorite activities was collecting agates

FIGURE 1
Norman and Marie Huber at "Summerland" camp, Mount Rainier, 1921. N. *King Huber collection.*

on its beaches, an early indication of an interest in "stones." Even without mountains my parents continued outdoor activities, and summer weekends were spent on excursions into the woods looking for wildflowers to transplant to my mother's garden. I developed a broad interest in nature, accumulating an extensive insect collection as well as stones. Graduation from high school and service overseas in World War II ended this phase of my life.

While I was overseas my parents moved to Lancaster, Pennsylvania, which became my new home upon my discharge. I needed to put down roots and consequently enrolled in a local college, Franklin & Marshall (F&M). For reasons not clear even to me, I decided to major in physics, but in the middle of my second year a colleague persuaded me to enroll in a geology course, which changed my whole outlook. I discovered that geology had three things going for it: first, the subject was absolutely fascinating; second, I was having trouble with differential equations, which geology did not seem to require; and, third, geology involved the out-of-doors!

The summer following my junior year I enrolled in a geology field course sponsored by Northwestern University. It consisted of a six-week excursion by canoe in what is now the Boundary Waters National Canoe Area in northern Minnesota. This allowed me to see some of my old stomping grounds in a new light and reinforced my decision to major in geology. In my senior year at F&M I dated the Geology Department's secretary, Martha Barr, who went by the nickname of "Nan," and decided to enter Graduate School at Northwestern.

Nan and I were married in 1951 at the end of my first year at Northwestern and honeymooned on the Mesabi Iron Range in northern Minnesota where I worked for an iron-mining company. The following two summers I was employed by the USGS on Michigan's Upper Peninsula, while Nan stayed behind and worked to help support my studies financially. After completing course work at Northwestern for my Ph.D. degree, we moved to Iron Mountain, Michigan, to start my full-time career with the USGS in 1954.

Soon after our 1955 arrival in California we were off to Auburn, a historic town in the Sierra Nevada foothills where I was to assist senior geologist Lorin Clark in geologic studies of the American River country. Part of this involved difficult traverses down nearly inaccessible canyons of the American River where the rock exposures were better for study than in the forested uplands. It was like mountains upside down. After spending the winter plotting data from our summer's activities and writing reports, I received a new assignment, working with geologist Dean Rinehart mapping the geology of the Devils Postpile quadrangle, an area of about 240 square miles in the Sierra Nevada just east of Yosemite National Park. The bulk of this quadrangle lies just west of the Sierran crest or drainage divide, and the only road penetrating any distance in from the east was a gravel road to the Devils Postpile National Monument and pack-stations along the Middle Fork San Joaquin River. Over three summers we became quite familiar with those packers, and they with us. For our fourth summer, we were located near the western margin of the quadrangle with access from that side. By the end of that summer I almost began to consider myself a "mountain man."

The Postpile project opened the magnificence of the High Sierra to me in one of its most spectacular regions. The wilderness character of most of the area also introduced me to new methods of field operations utilizing horses and pack mules—I was now truly following in my parents, mountaineering footsteps (and saddles). Fortunately, the animals were handled by a packer who would shepherd us to a base camp in the mountain range and leave us to work for a week or so until returning to pick us up. In addition to assisting Dean geologically I wound up being camp cook. In those days the modern lightweight camp foods were not available so we used a lot of canned goods. This made us thankful for the mules, but every time we hired a new packer we had to educate him that we geologists would replace our food with rocks and that it would require as many mules to pack us back out as it had to pack us in.

I gloried in the magnificent setting in which I found myself, with crystal-clear lakes set in glacial basins adjacent to massive Mounts Ritter and Banner and the jagged Minarets of the Ritter Range (FIG. 2)—all of this and absolutely fascinating geology. This was an area with abundant and diverse metamorphic rocks, whose character we were to decipher, but less of the granitic rock that

FIGURE 2
King's introduction to the Minarets in the Ritter Range, 1956.
N. King Huber collection.

makes up the bulk of the Sierra Nevada. Thus, unlike most Sierra visitors, we were challenged geologically as well as logistically.

It was not all work and no play and I quickly became an ardent trout fisherman. Dean and I spent many an evening or rainy day stocking our larder with fresh trout—once we even built a small smoke-oven. Sitting around the campfire with Dean playing his recorder, or lying on an outcrop watching shooting stars, were new and enjoyable experiences for me. My eyes opened to a whole new world.

The Geologic Division of the USGS has long had a policy of rotating its scientists through administrative and staff positions, and I was no exception. In the fall of 1959 my family and I, now with two young sons, transferred to the Survey's Washington, D. C. Headquarters, where I was to serve as staff assistant to the Associate Director. This sometimes exasperating, but educational, experience provided insight into the inner workings of the organization and an appreciation for its overall impact as an objective and credible scientific agency. My "servitude" also had a positive result; my proposal to return to Sierran studies was accepted.

Upon our return to California in the fall of 1961, I began geologic mapping in the Shuteye Peak quadrangle, southwest of the Devils Postpile quadrangle and just south of Yosemite National Park. I had barely started when one of two Navy jet fighters hop-scotching across the Sierra crashed nearly vertically into a grove of trees adjacent to a high meadow at 7000 ft elevation. When a search-and-rescue team of Marines arrived, I had the only topographic maps of the area. Using the angle of broken trees, I plotted the likely trajectory of the incoming plane. The Marines eventually found the body of the pilot along this trajectory on a ridge more than 1000 feet higher than the meadow. He had ejected, but too low to the ground for his chute to fully open.

A very rough jeep trail led to the site, and, because my Government jeep was the only four-wheel drive available, I drove upward with the Deputy Sheriff/Coroner hanging on but ready to jump if necessary. Hauling down a body-bag was not the best way to start a new project, but during this event I got to know Johnny Jones and his wife who ran a pack station at nearby Muglers Meadow. His station had the only outbuilding with tiled showers and flush toilets east of the resort community of Bass Lake, many miles over rough dirt roads to the west in the Sierra foothills. These facilities would be a luxury for my family the following year.

The summer of 1962 found a small government house trailer parked at Johnny Jones' pack station. The trailer only had sleeping

FIGURE 3
Steve, Rick, King, and Nan at Johnny Jones' Pack Station, Muglers Meadow, 1963. N. King Huber collection.

room for my wife Nan and younger son Rick, then three years old. Our five-year-old son Steve and I slept in one tent, and my field assistant in another tent. This was all at the edge of Muglers Meadow with a small brook meandering through it. What a site for little kids to learn all about bugs and frogs and cow pies! For three consecutive summers we camped there (FIG. 3) and at various Forest Service campgrounds, where weekend campers would kindly leave goodies at our government trailer when they left for home. Although not a born camper, Nan cheerfully put up with this nomadic stuff for the benefit of the kids, who could not have been happier.

During this period our geologic map of the Devils Postpile quadrangle was published, and I was working to complete the Shuteye Peak geologic map when I received a call from Chief Geologist Hal James, who had earlier been my boss on the Michigan iron project. He had enjoyed my recent layman's article on the Devils Postpile and asked me if I was interested in mapping the geology of Isle Royale National Park in Lake Superior. This was back home to me and my immediate answer was, "when do we go!" Although leaving the mountains was not easy, I had always admired Isle Royale from afar as I collected agates on Lake Superior beaches as a youth.

This was another new adventure for my family. Getting to the island was an all-day trip from Michigan's Upper Peninsula on a Park Service vessel, the magnificent *Ranger III*, with a capacity of 110 passengers and miscellaneous cargo for the island's needs. Isle Royale is an archipelago with park headquarters located on one of the smaller islets. Our boys were seven and nine that first year, and our family was provided with an apartment right on the shore of Lake Superior. We spent four glorious summers there while I mapped the island's geology. The main island is about 50 miles long, averaging about 6 miles wide. It does not have a single road on it, and my assistant and I would go off by boat for a week or so at a time to do field work (FIG. 4). Except for visitor facilities at each end of the island the area is completely wilderness. Unlike the Sierra wilderness it is a north-woods environment with very low relief. The highest point on the island is only 800 feet above lake level and the heavily wooded terrain makes rock outcrops difficult to locate. But such an environment has its own fascinating flora and fauna, highlighted by the wolf/moose, predator/prey relationship—not to mention mosquitoes.

FIGURE 4
Isle Royale montage. All youngsters were required to wear lifejackets when outside their residence because of dangers at the main dock. Small boat is the family transportation. Boat at upper right is the Park Service's "D. J. Tobin," used by King in his geologic field work. Float-plane was used by King for aerial reconnaissance trying to locate possible rock outcrops in wooded areas. N. King Huber collection.

The Isle Royale project was also an educational experience for our two boys. During those four years we drove eight different routes between California and Michigan, hitting many national parks on the way, even if only for a brief visit. One year this included a "grand circle" tour southeast through San Antonio for the Hemisfair, back on the Trans-Canada Highway to British Columbia, and then south to Olympic National Park.

In addition to several technical reports, the Isle Royale project resulted in a geologic map of the island and a layman-oriented book, *The Geologic Story of Isle Royale National Park*. Writing this book was a valuable experience that would aid my later efforts to interpret the geology of the Sierra Nevada for the general public. There was also a surprising connection between my work on Isle Royale and the Sierra. The first geologic map of the island, published in 1850, had as coauthor one Josiah D. Whitney, who as State Geologist would later lead the California Geological Survey in its Sierran studies and thus be my geologic predecessor in both locales.

Back in 1959 when Dean Rinehart and I were mapping in the Devils Postpile quadrangle, we spent a considerable amount of time packed in to the remote upper reaches of the North Fork San Joaquin River. We had the place virtually to ourselves as it was a dead end with no through trails, and we thoroughly enjoyed it. Vowing to return with our children, we did so in 1972 when Dean, his wife and teenage daughter, and I and my two sons hiked back in to find mobs of people, including some that we knew. One evening a chap and his young daughter entered our camp. He had heard that we were geologists and asked if we knew King Huber with the USGS in Menlo Park. Flabbergasted, I had never met this Yale biology professor but had corresponded with him regarding the geological impact of moose on their vegetative fodder on Isle Royale. He had done his Ph.D. research in the North Fork basin and had vowed to bring his daughter there when she was old enough.

After completing my Isle Royale reports, I returned to the High Sierra for a new venture. My assignment was an evaluation of the mineral-resource potential of the Minarets Wilderness Area as mandated by the Wilderness Act of 1964. Much of this area, since renamed the Ansel Adams Wilderness, was within the Devils Postpile quadrangle, so I was back on familiar ground. The work this time was different, however. The geology had already been mapped, mostly by Dean Rinehart and me, so the chief task was to examine the many mining claims and to sample stream sediments for analysis to detect anomalous concentrations of metallic elements. My three field assistants and I spent two summers completing this task. Equally important for me, by paying for extra mules and horses, my two teenage sons were able to spend two full summers with me in the mountains, an experience that they will never forget. My biggest problem was limiting the number of fish they could catch. As delicious as fresh trout is, my crew did not want fish every meal.

Upon completion of the Mineral Resources report, I undertook a study of the amount and timing of the uplift of the Sierra Nevada mountain range. This analysis used dated remnants of volcanic flows within the canyon of the San Joaquin River, which once flowed across the Sierra from the region now occupied by the Mono Lake basin and Long Valley on the east side of the range. I later made a similar study of the evolution of the Tuolumne River heading in Yosemite National Park. Returning to geologic mapping,

FIGURE 5
King leads discussion of Yosemite geology with a group of park interpreters in Yosemite Valley. N. King Huber collection.

I completed geologic maps of the Pinecrest quadrangle, which includes the northwest corner of Yosemite National Park, and the Dardanelles Cone quadrangle immediately to the north. My field assistant in the Pinecrest quadrangle, Barbara John, and I were featured in the USGS Geologic Division segment of the *Smithsonian Magazine* issue on the 100th anniversary of the USGS in 1979.

The year 1981 saw the initiation of our formal Yosemite project, which is described in the accompanying article on "Interpreting Yosemite Geology." My Yosemite experience was personally very satisfying to me, and I am proud of the geologic map, book, and other products that resulted. Equally satisfying were the many friendships that evolved with the Yosemite staff and associated organizations (FIG. 5).

Along the way, I had the privilege of conducting field excursions for Congressional-staff members with USGS responsibilities and for President Reagan's Science Advisor.

The fall of 1990 saw the Centennial Celebration of Yosemite as a National Park. The Director of the USGS was involved with Congressional matters, and I was consequently designated as the official USGS representative. My participation also involved presenting a paper at a centennial symposium and leading geologic field trips in Yosemite Valley (FIG. 6).

I joined the Wilderness Historic Resources Survey with historian Jim Snyder, packer Bob Barrett, cook and artist Jim Murphy and other volunteer helpers in 1988–89. In addition to the wonderful camaraderie, this permitted me to visit geologic sites that I had previously seen only in photographs. Included were "The Slide," a massive rockslide in Slide Canyon in northern Yosemite, and the volcanic-filled, ancient channel of the Tuolumne River on Rancheria Mountain. My articles on these geologic features follow in this book.

In the spring of 1994, I officially retired from the USGS after 40 years of service and a lot of fun. Through a Scientist Emeritus appointment, the Survey continued to provide office space for me, and I took on new Yosemite projects. After my colleague Clyde Wahrhaftig died I finished compiling and saw through to publication his *Geologic Map of the Tower Peak Quadrangle* in northern Yosemite National Park. Clyde had a heart condition that required open-heart surgery. Concerned that he would not live to finish several cherished projects, his will established a trust fund to aid in their completion by others. Funneled through the Yosemite Association, these funds supported the publication of his Tower Peak geologic map.

I also started to write a series of articles for the quarterly journal of the Yosemite Association to make some of Yosemite's complex geology more accessible to a broader public. Those articles with revisions make up much of this volume. I hope that I have been a worthy successor to François Matthes in bringing the geologic story of Yosemite to the public. I could not have asked for a more welcome opportunity to do so.

FIGURE 6
King explains field work in glacial geology above Rodgers Lake in 1989. *Dwight Barnes photo.*

ACKNOWLEDGMENTS

The articles that make up this volume were not written without considerable help from colleagues and friends who improved both technical and editorial aspects. USGS colleagues providing reviews include Bob Christiansen, Malcolm Clark, Steve Ellen, Russ Evarts, Bob Jachens, Ron Kistler, Jim Moore, Dick Pike, Dean Rinehart, and Rowland Tabor. Editorial help was also provided by my neighbor, Ken Arendt.

N. KING HUBER
Menlo Park, CA, 2006

Incomparable Valley

The Geologic Story of Yosemite Valley

BY N. KING HUBER

YOSEMITE COUNTRY

For its towering cliffs, spectacular waterfalls, granite domes and spires, glacially-sculpted and polished rock, and beautiful alpine scenery, Yosemite National Park is world famous. Nowhere else are all these exceptional features so well displayed *and* so easily accessible. Artists, writers, tourists, and geologists flock to Yosemite—and marvel at its natural wonders. Yosemite Valley itself is deeply carved into the gently sloping western flank of the Sierra Nevada, the longest, the highest, and the grandest single mountain range in the United States outside of Alaska. And although other valleys with similarities exist, there is but one Yosemite Valley, the "Incomparable Valley" of John Muir, California's most famous naturalist.

FIGURE 1
Bird's-eye view of Yosemite Valley and the High Sierra, looking eastward up the valley. USGS.

A SCENE OF WORLDWIDE FAME

Simply stated, Yosemite Valley, only 7 miles long and nearly 1 mile wide, is a flat-floored, widened part of the canyon of the Merced River. But this broad rock-hewn trough with roughly parallel sides is boldly sculptured and ornamented with silvery cataracts.

FIGURE 2
El Capitan's bold profile faces Cathedral Rocks and Bridalveil Fall on the right. N. *King Huber photo.*

FIGURE 3
Leaning Tower and the west flank of Cathedral Rocks frame Bridalveil Fall. N. *King Huber photo.*

From the valley floor at an elevation of 4,000 feet, the magnificent cliffs rise 3,000 to 4,000 feet higher to forested uplands on either side (FIG. 1).

Once you enter Yosemite Valley its grandeur is overwhelming. Looking eastward up the valley from its lower end you are struck by the immensity of the sheer profile of El Capitan, the most majestic cliff in the valley (FIG. 2). Projecting boldly from the north wall, its top rises 3,000 feet above the valley floor. Directly opposite stand the Cathedral Rocks, over 2,500 feet high, which also jut into the valley. Between the west end of this promontory and the Leaning Tower, Bridalveil Fall leaps 620 feet, its abundant spray commonly suffused with rainbows (FIG. 3).

Eastward beyond the narrows at El Capitan and Cathedral Rocks, the valley abruptly widens, and in an embayment on the south are the Cathedral Spires, among the frailest rock shafts in the valley (FIG. 4). On the north are the Three Brothers (FIG. 5), whose gabled summits rise one above another, all built architecturally on

FIGURE 4
The Cathedral Spires project skyward to the left of Cathedral Rocks.

the same angle. The highest, known as Eagle Peak, rises nearly 3,800 feet above the valley floor. Across the valley stands Sentinel Rock, a finely modeled obelisk with a pointed top (FIG. 6).

A mile farther up the valley, on the north side, are Yosemite Falls, dramatically booming among clouds of mist during the spring and early summer snowmelt (FIG. 7). The Upper Fall, 1,430 feet high, would alone make any valley famous; it is the highest unbroken leap of water on the continent. The Lower Fall, which descends 320 feet, seems insignificant by comparison, yet it is twice as high as Niagara Falls. The entire chain of falls and intermediate cascades drops 2,425 feet. Ribbon Fall, west of El Capitan, descends 1,612 feet, but it is

FIGURE 5
The Three Brothers, whose angled summits slope westward. N. King Huber photo.

FIGURE 6
Sentinel Rock's Pinnacle juts boldly upward. N. King Huber photo.

FIGURE 7
Upper Yosemite Fall now leaps from the hanging valley of Yosemite Creek. In the not-too-distant geologic past its water cascaded down through the prominent ravine to the left (see figure 13). The tree-covered bench at middle left marks the approximate upper limit of the Tioga-age glacier in Yosemite Valley. N. King Huber photo.

confined in a sheer-walled recess and does not make a clear leap throughout.

Farther up the valley, on the north side, are the Royal Arches, sculptured one within another into an inclined rock wall that rises 1,500 feet (FIG. 8). An enormous natural pillar, the Washington Column, flanks them on the right, and above them rises a smoothly curving, helmet-shaped knob of granite called North Dome. Facing the Royal Arches on the south wall, stands Glacier Point providing a matchless view of the valley from its summit, which stands 3,200 feet above the valley floor.

At the head of the valley, as if on a pedestal, stands Half Dome, the most colossal and recognizable rock monument in the Sierra Nevada, smoothly rounded on three sides and a sheer vertical face on the fourth (FIG. 9). From its summit, over 4,800 feet above the valley, you look southeast into Little Yosemite Valley, which is broad floored and has granite walls more gently sloping than in its larger namesake. From Little Yosemite's western portal, guarded by Liberty Cap, the Merced River descends by a giant stairway, making two magnificent waterfalls, Nevada Fall, dropping 594 feet, and Vernal Fall, dropping 317 feet. Looking northward from Half Dome's summit, the view is into Tenaya Canyon, a chasm as profound as Yosemite Valley itself, yet the pathway of only a small brook. To the northeast, Clouds Rest, the loftiest summit in the vicinity of Yosemite Valley, rises to 9,926 feet; beyond, spreads the vast panorama of the High Sierra.

The present Yosemite Valley is the result of many different geologic processes operating over an incomprehensible length of time measured in millions of years. These processes are by themselves not unique, but their unrivaled interaction has created this "Incomparable Valley." The accumulated observations, studies, and interpretations by many individuals through the years allow us to reconstruct much of the valley's geologic history, adding to our appreciation of its scenic majesty.

FIGURE 8
The Royal Arches are flanked by Washington Column and surmounted by North Dome. *N. King Huber photo.*

FIGURE 9
Half Dome's cleaved face dominates the east end of Yosemite Valley. *N. King Huber photo.*

ROCK, THE SCULPTOR'S MEDIUM

For any form of sculpture, whether of finely-chiseled statues or massive landforms, the resulting product is strongly dependent on the nature of the material being worked upon. For Yosemite Valley that material is granite. Indeed, granite forms the bedrock of much of the Sierra Nevada, including most of Yosemite National

Park. Granite, in the broad sense of the term (*granitic rock*), is a rock with a salt-and-pepper appearance due to random distribution of light and dark minerals. The mineral grains are generally coarse enough to be individually visible to the naked eye.

Throughout the park granitic rock varies considerably in the relative proportions of the individual light and dark minerals, and these compositional differences are represented by a variety of specific names, such as granodiorite and tonalite, in addition to "true" granite as defined by geologists.

From a distance, all of Yosemite Valley's granitic rock looks the same. But it actually consists of individual rock bodies, each with their own characteristic mineral composition and texture, that is, the coarseness of their crystals and uniformity or variation in grain size. All of these variations, in turn, affect the rock's resistance to abrasion, fracturing, and weathering, all important to the sculptor and the end product. The imposing cliffs of El Capitan and Cathedral Rocks, for example, are composed of a particularly tough and resistant variety of granitic rock.

THE ROLE OF JOINTS

The bedrock structures having the greatest effects on Yosemite's landform development are *joints*. Although granitic rock is unbroken on a small scale, on a larger scale the rock is broken by joints, which are more or less planar cracks commonly found as sets of parallel fractures in the rock. Regional-scale joints commonly determine the orientation of major features of the landscape, such as the planar face of Half Dome, the series of parallel cliffs at Cathedral Rocks, and the westward sloping faces of the Three Brothers. In contrast, smaller, outcrop-scale joints determine the ease with which rock breaks and erodes. Joints are of overwhelming influence on landform development in granitic terrain because they form greatly contrasting zones of weakness in otherwise homogenous, erosion-resistant rock and allow access for water and air to enter and aid in the weathering and disintegration of rock.

The type of jointing that most influences the form of Yosemite's landmarks, however, is the broad, onion-shell-like *sheet jointing* formed by a process referred to as *exfoliation*. Granitic rocks originate at considerable depth within the Earth while under great pressure from overlying rock that may be miles thick. As the overlying rock erodes away, the decrease in the pressure that once confined the granitic rock causes it to expand toward the Earth's surface. When the outer expanding zone exceeds the strength of the rock, it cracks away from less expanded rock beneath and bursts loose as a sheet; subsequent cracks release successive layers of expanding rock. Because the expansion that forms these sheets takes place perpendicular to the local surface, the shape of sheets generally reflects topography, with curved surfaces following hill and valley. Sheet joints also tend to parallel the walls of canyons and cliffs that may appear to be unbroken monoliths, such as El Capitan, but which may have fractures behind and parallel to the cliff face. The curved upper surfaces of North Dome and Half Dome, and the undulating surface of Clouds Rest, are magnificent examples of sheet joints or sheeting (FIGS. 8, 9).

Admiring Yosemite Valley's intricately sculptured walls as they appear today and knowing something about the granitic rock from which they were carved, we can look back in time to speculate on the evolution of the valley's formation—its geologic history.

A STORY THAT BEGAN MILLIONS OF YEARS AGO

The last touches to Yosemite Valley's architecture were applied relatively recently, geologically speaking. But the rock from which the valley is carved originated mainly during the Cretaceous period, about 100 million years ago, when dinosaurs roamed the Earth. At that time molten rock, *magma*, generated deep within the Earth, rose upward within the Earth's crust, or upper layer, and crystallized far beneath the surface to form granitic rock along a linear belt that was to become the future Sierra Nevada. The granitic terrain that makes up the Sierra, once thought to have only local variations in one huge mass of rock, is actually made up of a mosaic of individual rock bodies that formed from repeated intrusions of magma over many millions of years.

Some of the magma broke through to the surface, building a string of volcanoes atop hidden granitic roots, and we can perhaps envision an ancient majestic mountain range somewhat like the modern Cascade Range along the coast of our Pacific Northwest. Because of the high elevation of this ancestral range, however, the volcanic and other rocks covering the granite were soon eroded away, and by Late Cretaceous time, about 70 million years ago, the granitic rocks became exposed at the Earth's surface. By middle Cenozoic time, a few tens of millions of years ago, so much of the upper part had been removed that in the vicinity of Yosemite the surface of the range had a low relief of only a few thousand feet.

Later, the continental crust east of the Sierra Nevada began to stretch in an east-west direction, developing into a series of north-south-trending valleys and mountain ranges. Through a combination of uplift of the Sierran block and down-dropping of the area to the east, the Yosemite region acquired a tilted-block aspect with a long, gentle slope westward to the Central Valley of California and a short, steep slope separating it from the country to the east.

AGENTS OF EROSION

Erosion, simply stated, is the removal of earth materials from high areas to lower areas, modifying the landscape in the process. Two agents of erosion are chiefly responsible for sculpting the present Yosemite landscape—flowing water and glacial ice: flowing water had the major role, and glacial ice added additional touches. The general Sierran landforms were all well established before glaciation, and the major stream drainage systems provided the avenues along which the glaciers would later follow. Some of the glacial modifications, however, were profound: the creation of alpine topography in the High Sierra, the rounding of many valleys from V-shape to U-shape, and the straightening of valleys in the process. Still another agent of erosion is simply gravity. The downslope movement of rock materials produces landslides and rockfalls. Although generally of local extent, such movement is important, particularly in mountainous terrain and on the over-steepened slopes in Yosemite Valley.

The Role of Flowing Water

The effectiveness of erosion by flowing water depends both on processes of weathering—the breakdown of larger rocks into smaller individual rock and mineral fragments that can be transported—and on stream volume and velocity, which determines the size and amount of material that can be transported. With the increasing late Cenozoic elevation of the Yosemite region, the major streams coursing down its western slope were rejuvenated and made more vigorous by their increased slopes. Under these

(a)   (b)

FIGURE 10
The role of flowing water. A few tens-of-millions of years ago the Yosemite area was a rolling surface of rounded hills and broad valleys with meandering streams (a). Before the onset of glaciation, more than a million years ago, the elevation of the range increased and streams incised deep canyons into the western flank of the range (b). *USGS.*

conditions the major streams cut canyons whose channels became progressively deepened relative to the upland areas between them, areas which even today retain comparatively moderate relief. The upper basins and middle reaches of the Merced and Tuolumne Rivers, for example, were later modified by glacial erosion, but initial canyon cutting was accomplished solely by the action of streams. Two sketches depict an artist's conception of the evolution of the Merced River canyon at the site of the future Yosemite Valley before the onset of glaciation (FIG. 10).

The Role of Glaciers

The Yosemite landscape as we see it today strongly reflects the dynamic influence of flowing ice that long ago covered much of its higher regions. Geologists are still uncertain how many times ice mantled Yosemite, but at least three major glaciations have been well documented elsewhere in the Sierra Nevada. In the higher country, icefields covered extensive areas, except for the higher ridges and peaks. Lower down the western slope, at middle elevations, glacial tongues were confined to pre-existing river canyons, such as those of the Merced and Tuolumne Rivers. Thus our focus will be on the nature and activities of these valley glaciers, particularly as they apply to Yosemite Valley, and Hetch Hetchy Valley some 15 miles to the north, but remembering that the valley glaciers derive their flowing ice from icefields higher in the range.

In contrast to the sinuous V-shaped valleys of normal streams in unglaciated mountainous terrain, glaciated valleys tend to be straighter and have U-shaped profiles. Whereas a stream erodes the outsides of bends preferentially and makes its course more sinuous, glacial erosive force is concentrated on the insides of bends, removing the protruding spurs of the original stream valley and leaving a wider, straighter valley.

The resulting modification, in detail, depends on the nature and structural integrity of the bedrock over which the glacier is flowing. For granitic bedrock, the dominant structure of concern is jointing, which controls the ease of removal of rock that is otherwise highly resistant to glacial erosion.

YOSEMITE VALLEY AND ITS GLACIERS

Yosemite Valley has often been referred to as a "classic" glacial valley. But what glacially-derived attributes does it display to deserve that designation? A glacier tends to straighten a valley and smooth its walls as it grinds past them. But the walls of Yosemite Valley are extremely ragged, with many pinnacles and spires projecting upward from them—Leaning Tower, Cathedral Spires, Sentinel Rock, and Lost Arrow stand out strikingly. All of the waterfalls and lesser cascades along the sides of the valley are ensconced in alcoves, except for Upper Yosemite Fall, whose story will be told in upcoming paragraphs. Eagle Creek and Indian Canyon Creek actually issue from deep ravines. All of these seemingly anomalous features would doubtless be obliterated by a glacier that filled the valley to its brim. And yet glacial erratics—boulders transported and deposited by a glacier—are found scattered above the valley's rim, telling us that a glacier indeed once filled the valley to its brim. How can we

FIGURE 11
Hetch Hetchy Valley before damming. Note the absence of Yosemite-like pinnacles. *Yosemite Research Library.*

explain this anomalous appearance if the valley was indeed shaped by a glacier? The anomaly is even more apparent if we compare Yosemite Valley with another "glaciated" valley of about the same size and elevation, Hetch Hetchy Valley, which has comparatively smooth walls and an absence of pinnacles and spires (FIG. 11).

Little doubt exists that Yosemite Valley indeed represents a profound, glacially-driven modification of the Merced River canyon, as no other erosive agent could have accomplished such excavation. A glacier filling the valley to its rim created the basic broad shape of the valley and gouged out a deep bedrock basin whose bottom locally, in its eastern part, lies more than 1,000 feet below the present valley floor (FIG. 12A). That glacial episode was named the El Portal glaciation by François Matthes in his monumental Yosemite study, because he thought that its glacier advanced down the Merced canyon to near the community of El Portal, some 10 miles downstream from Yosemite Valley proper. Today we correlate that glaciation with the Sherwin glaciation, defined from studies along the east side of the Sierra Nevada, and which name is now in general use. The Sherwin was the most extensive, and longest-lived, glaciation documented in the Sierra. It may have lasted almost 300,000 years and ended about 1 million years ago. A Sherwin-age glacier was almost surely responsible for the major excavation and shaping of Yosemite Valley within the Merced River canyon.

Later glaciations in the Sierra Nevada were of lesser areal extent and briefer than the Sherwin. The best documented are the Tahoe and Tioga glaciations, which probably peaked about 130,000 and 20,000 years ago, respectively; together they are equivalent to Matthes' "Wisconsin" glacial stage, which he did not subdivide. The last glacier in Yosemite Valley—Tioga in age—advanced only as far as Bridalveil Meadow (FIG. 12B). At this location the forward

(a)

(b)

FIGURE 12
Glaciers large and small. Sketches of Yosemite Valley area, showing extent of valley-filling Sherwin glacier (a) and lesser extent of Tioga glacier (b). USGS.

movement of the glacier was balanced by the melting of ice at its front, or terminus. A "terminal" end moraine—a low ridge crossing the valley—was constructed with rock debris transported by the glacier and deposited at its terminus. The extent of the earlier Tahoe-age glacier in the valley is uncertain, but evidence elsewhere in the Sierra suggests that it probably would have been somewhat longer than the Tioga. Nevertheless, since the original excavation of Yosemite Valley by a Sherwin-age glacier, no subsequent glacier has filled the valley to its rim, a conclusion that has important consequences for the scenery.

From its terminus at Bridalveil Meadow, the ice surface of the Tioga glacier would have sloped upward toward the east end of the valley with the ice reaching a thickness of perhaps a little over 1,000 feet at Columbia Rock west of Yosemite Falls, 1,500 feet at Washington Column, and 2,000 feet in Tenaya Canyon below Basket Dome, as reconstructed by Matthes. Thus the Tioga and similar Tahoe glaciers could do very little to further modify or smooth the walls of Yosemite Valley. Above the ice surface of those glaciers, the valley walls have had a million years to weather: joints widened, rock fractured and crumbled, and waterfalls and cascades eroded back into alcoves and ravines. Thus the pinnacles and spires that seem so anomalous for a glacial valley actually had a million years to form and, being above the level of later glaciers, remain to amaze us today. In Tenaya Canyon, Tioga ice was thicker and reached farther up the walls, smoothing them and removing irregularities; no pinnacles and spires are found there.

Hetch Hetchy Valley on the Tuolumne River, otherwise similar to Yosemite Valley, has comparatively smooth walls and an absence of pinnacles and spires (FIG. 11). There the Tioga glacier was also less extensive than the Sherwin, but unlike the glacier in Yosemite Valley, the Tioga glacier filled Hetch Hetchy to the rim. Thus Hetch Hetchy Valley's walls were being scraped and debris was removed from the valley with each glaciation, including the last. The greater extent of the glacier in Hetch Hetchy can be attributed to the fact that the drainage basin of the Tuolumne River above Hetch Hetchy is more than three times as large as that of the Merced River above Yosemite Valley. As a result, the much larger icefield feeding the Tuolumne glacier was able to provide the necessary volume of ice to fill Hetch Hetchy even though the Tioga glaciation was regionally less extensive than the Sherwin. The smaller Merced icefield was unable to provide sufficient ice to fill Yosemite Valley, even though supplemented by ice from a part of the Tuolumne drainage that flowed southwest over a low pass into Tenaya Canyon.

LEAPING FALLS AND HANGING VALLEYS

Waterfalls leaping out from a valley's walls far above the valley floor have long been considered evidence of a glacial origin for that valley. The enormous Sherwin-age glacier that shaped Yosemite Valley was able to excavate the central chasm to a greater depth than smaller glaciers in side-entering tributaries. The result was that some of the side valleys were left "hanging" with waterfalls at their brinks. Since Sherwin time, most of the tributaries have eroded their channels back into the walls to leave little more than steep ravines with minor falls interrupted by chains of cascades, such as those at Sentinel Fall. Bridalveil Fall is an exception, although it also has receded back into an alcove from its original position further out on the valley wall.

In contrast to Bridalveil Fall, Upper Yosemite Fall, although now leaping from a hanging valley, had a very different origin. Yosemite Creek is the largest stream flowing into the north side of Yosemite Valley and probably entered the Merced River canyon through a steep side canyon before glaciation. After the Sherwin glacier deepened Yosemite Valley, Yosemite Creek continued to enter the main valley through that ravine just west of the site of the present fall (FIG. 7). Matthes recognized this and described "what appears to be an old stream channel" just to the west of the present channel (FIG. 13). At that time the site of the present Upper Fall hosted only a minor ephemeral fall of short duration during spring runoff.

FIGURE 13

Yosemite Creek, then and now. Today Yosemite Creek (heavy line) flows over the valley rim to create Upper Yosemite Fall. Before its diversion about 130,000 years ago, Yosemite Creek flowed down an older channel just to the west (heavy dashed line), from which it cascaded down through the steep ravine that is now the route of the Yosemite Falls trail. Also shown is the complex of glacial moraines as mapped by Matthes (1930). USGS.

Matthes did not speculate on how or when Yosemite Creek was diverted from that old channel into its present channel to create its Upper Fall. He did, however, map a morainal complex that he attributed to his "Wisconsin-age" glacier that flowed down Yosemite Creek, but stopped about one-half mile short of the rim of Yosemite Valley itself. A plausible explanation for the diversion of Yosemite Creek into its present channel is that the stream was temporarily blocked by glacial deposits and had to find a new way through the intricate complex of nested moraines. As his Wisconsin glacial stage includes both Tahoe and Tioga glaciations, Upper Yosemite Fall, with its "newly" hanging valley, can be little more than 130,000 years old. And what a spectacular addition to Yosemite Valley's architectural wonders it is!

YOSEMITE VALLEY'S GLACIAL FINALE

While the Tioga glacier was constructing its terminal moraine at Bridalveil Meadow, the climate apparently warmed slightly. The ice at the front of the glacier began to melt faster than the ice was moving forward, and the ice front, or "snout," of the still-flowing glacier began to "retreat" up the valley. The climate cooled again; the ice front paused and temporarily stabilized just west of El Capitan Meadow. Here the glacier began to construct a new moraine, known as a "recessional" moraine because the glacier had receded from its terminal position. It remained at this location longer than it had at Bridalveil Meadow and the resultant El Capitan Moraine is larger in both volume and height. Eventually, the climate warmed abruptly, and the Merced glacier's snout retreated toward the head of the valley with no more recessional pauses, probably leaving Yosemite Valley by 15,000 years ago.

When the Tioga-age glacier departed from Yosemite Valley it left behind a lake, which Matthes christened *Lake Yosemite* (FIG. 14). It is likely that the advancing Tioga glacier had excavated some of the pre-existing valley fill east of the El Capitan Moraine, creating a shallow lake basin. The lake was in part dammed by this moraine, with the Merced River flowing over a low spillway through the moraine near the south valley wall. As the separate arms of the

FIGURE 14
Glacial Lake Yosemite filled the basin left upstream from the El Capitan moraine after the retreat of Tioga ice from Yosemite Valley. USGS.

Tioga glacier retreated up the Merced and Tenaya canyons, the melt-water-swollen, debris-laden rivers issuing from their snouts delivered large quantities of sediment to the lake basin. The lake was soon filled in with this sediment, creating the relatively level valley floor we see today. The resulting gentle slope allowed the Merced River to develop a sinuous meander pattern across this broad flood plain. A low-gradient, meandering stream is particularly susceptible to over-bank flow during high water, and its flood plain is naturally destined for periodic flooding.

THE ROCKS COME TUMBLING DOWN

The Tioga-age glacier did little to modify Yosemite Valley other than to remove pre-existing talus at the base of cliffs east of Bridalveil Meadow; all of the talus there now has accumulated in the last 15,000 years or so since the Tioga glacier departed. In contrast, the enormous talus slope west of El Capitan, known as the Rockslides, escaped the reach of the Tioga glacier. For the past million years, then, the rock walls of the valley that remained above the ice-level of the smaller post-Sherwin glaciers have weathered, joints have been enlarged, and rock has spalled off to form the very irregularly sculptured surface we see today. This geologic history provides the setting for abundant rockfalls. Every significant historical rockfall in Yosemite Valley has originated in vulnerable fractured rock from above the level of the Tioga glacier. Some rockfalls are quite large, but most are relatively small and gradually build up a cone of debris below the most active sites. Thus the size of a debris cone can reflect the volume or the frequency of individual falls, or a combination. The shattered rock high on the east side of Middle Brother provides material for a debris cone at one of the most active rockfall sites in the valley. Given the nature of the geologic setting, it is inevitable that such rockfalls will continue as part of ongoing geologic processes.

EPILOGUE

The geologic story of Yosemite Valley, as presented here, describes our present understanding of the interplay of various geologic processes that contributed to the valley's creation. But this is not the last word. We are still learning about these various processes and their effects on the evolution of the valley. And, indeed, these geologic processes are far from finished. We are seeing only a brief period in time in the landscape's ongoing history. Dynamic geologic processes will continue to change the many faces of this "Incomparable Valley."

ACKNOWLEDGMENTS

This text was originally prepared to be printed on the reverse side of the Topographic Map of Yosemite Valley published by the USGS. There it would replace an earlier text written by François Matthes in 1922. Not revised since 1938, Matthes' text does not reflect modern geologic concepts, developed in part with new technology, including isotopic dating methods. Matthes' descriptive materials that resulted from his pioneering Yosemite geologic studies, however, remain as vibrant as ever and the opening sections of the present text borrow from, and hopefully echo, the exuberance of his eloquent prose.

Exploring Yosemite Geology

Interpreting Yosemite Geology
A Historical Perspective

BY N. KING HUBER

INTRODUCTION

Through the years many individuals have contributed to understanding Yosemite geology, but here I only highlight the roles of two government organizations, the original California Geological Survey and the United States Geological Survey. Both groups hosted pioneering scientists whose contributions greatly advanced our geologic knowledge of the Yosemite region. Here some of their activities, observations, and conclusions are recounted, and the connections between the two organizations are examined. Clarence King, starting as an unpaid volunteer with the California Survey and finishing as the U.S. Geological Survey's first Director, is the thread that ties things together.

THE CALIFORNIA GEOLOGICAL SURVEY

The original California Geological Survey was created by the State Legislature in 1860. Renowned geologist and Harvard University Professor Josiah D. Whitney was appointed as State Geologist and Survey Director (FIG. 1). Following studies in other parts of the state, the California Survey mounted its first geological expedition to the Yosemite region in 1863, returning again in

FIGURE 1
Members of the California Geological Survey, December 1863, from left to right: Chester Averill, William Gabb, William Ashburner, Josiah Whitney, Charles Hoffmann, Clarence King, and William Brewer. *Courtesy of the Bancroft Library (CA Geological Survey—POR 1), University of California, Berkeley.*

subsequent summers for several years. Its members, including Whitney himself, explored from Yosemite Valley to Tuolumne Meadows and the High Sierra Crest, in the process naming many prominent features, such as Mounts Hoffmann, Dana, Gibbs, Maclure, and Lyell. Mount Hoffmann was named for one of their own party; the others were prominent scientists, including three geologists.

In 1865 Whitney published a volume summarizing the work of the California Survey during its first four years, including the Yosemite results of 1863 and 1864.[1] While in the high country, Whitney and others in his crew had observed evidence of former glacial activity. He wrote that "the whole region about the head of the Upper Tuolumne is one of the finest in the state for studying the traces of the ancient glacier system of the Sierra Nevada. The valleys . . . exhibit abundant evidences of having, at no very remote period, been filled with an immense body of moving ice, which has everywhere rounded and polished the surface of the rocks up to the height of at least a thousand feet above the present level of the river at Soda Springs [in Tuolumne Meadows]"[2] (FIG. 2).

With respect to Yosemite Valley itself, Whitney reported the observations of Clarence King, one of the Survey members, of evidence for glaciers in the valley. King recognized that low ridges crossing the valley between Bridalveil Meadow and El Capitan Meadow were actually glacial moraines that marked the downstream extent of a former glacier. King did not go so far, however, as to conclude that the valley was actually carved by a glacier, but only that it was once occupied by a glacier at least a thousand feet thick. Whitney noted that "this moraine [near El Capitan] may have acted as a dam to retain the water within the valley, after the glacier had retreated to its upper end, and that it was while thus occupied by a lake that it was filled up with the comminuted materials arising from the grinding of the glaciers above, thus giving it its present nearly level surface."[3] This was a prescient observation, as the former existence of at least a small post-glacial lake has been substantiated by later workers.

Whitney was greatly impressed by "Cliffs absolutely vertical, like the upper portions of the Half Dome and El Capitan, and of such immense heights as these. "They are," he wrote, "so far as we know, to be seen nowhere else." This led him to propose that the valley resulted from a down-dropped fault block,[4] a conclusion that was to later become the focus of a dispute with naturalist John Muir. But that's another story.

In June 1864, President Lincoln signed the bill that granted Yosemite Valley and the Mariposa Grove of Big Trees to the State of California to "be held for public use, resort, and recreation for all time." Whitney and the California Geological Survey had more than an incidental connection to the Yosemite Grant. The bill was carried in the U. S. Congress by California Senator John Conness who, while in the State Senate in 1860, had introduced the act that created the California Survey. In appreciation, Whitney named a major peak on the Sierran crest north of Tioga Pass as Mount Conness for the Senator. A Board of Commissioners, which included the Governor of California and eight others, was created to manage the Yosemite Grant. Frederick Law Olmsted, one of the foremost landscape planners in the country but at that time manager of Frémont's former Mariposa Estate west of Yosemite, became acting chairman of the commission. Whitney and William Ashburner, a former member of the California Survey, were also appointed to the commission. Helping to promote the congressional bill was a set of pioneer photographer Carleton Watkins' mammoth 18″ × 22″ photographic views, which demonstrated the uniqueness of the Yosemite landscape more graphically than any words. Watkins was also to become the primary photographer for the California Survey.

Even before his appointment to the commission was formalized, Olmsted hired Clarence King and James Gardner, both of

FIGURE 2
"Glacier-polished Surfaces in Tuolumne Valley," a sketch by Charles F. Hoffmann from Whitney's *Geology*, 1865.

FIGURE 3
Topographic map of Yosemite Valley prepared by Clarence King and James Gardner, showing boundaries of the Yosemite Grant established by their survey in 1864, from Whitney, 1869.

the California Survey, to survey and define the boundaries of the Yosemite Grant as required under the federal act. The area of concern, as stipulated in the act, was "the Yosemite Valley, with its branches and spurs, in estimated length fifteen miles, and in average width, one mile back from the main edge of the precipice, on each side of the Valley."[5] In fall 1864, King and Gardner completed the required survey and produced the first topographic map of the valley upon which the Grant boundaries were defined (FIG. 3). The boundary lines were determined by connecting a series of prominent topographic highs around the valley from which the line-of-sight could be made with their survey instruments. Building upon this first map, over the next few years the California Survey produced additional maps, and the region that included Yosemite Valley gradually became better depicted geographically.

In 1868 Whitney published his *Yosemite Book*, which expanded on the Survey's earlier information and included a glowing description of both Yosemite Valley proper and what he referred to as the "High Sierra." This volume was produced in a very limited edition due to the inclusion of 28 photographic prints, 24 of them by Carleton Watkins, which had to be individually inserted into the book.[6] For the benefit of the touring public another edition with the same text, but with woodcuts to replace the photographs, was issued as The

Yosemite Guide-Book beginning in 1869 with several later editions.[7] In these volumes Whitney further emphasized his fault-block hypothesis for the origin of Yosemite Valley. More importantly, he recanted his earlier statements noting Clarence King's observations on glaciers in Yosemite Valley. In a later report Whitney reinforced this view saying that he was unable to find any glacial polish in Yosemite Valley,[8] which was not surprising since even today readily accessible exposures of glacial polish in the main valley are scarce. But his rejection of King's observations was still puzzling since Whitney did recognize the existence of glaciers in Hetch Hetchy Valley, which he otherwise compared favorably with its Yosemite counterpart (FIG. 4). "There is no doubt that the great [Tuolumne] glacier . . . found its way down the Tuolumne Canyon, and passed through the Hetch-Hetchy Valley. Within the Valley, the rocks are beautifully polished, up to at least 800 feet above the river. Indeed, it is probable that the glacier was much thicker than this; for, along the trail, near the south end of the Hetch-Hetchy, a moraine was observed at the elevation of fully 1,200 feet above the bottom of the Valley."[9] He thought the glacial differences between the two valleys resulted from the much larger volume of the icefield feeding the Tuolumne than that feeding the Merced River leading into Yosemite Valley.[10] Regardless of Whitney's errors on Yosemite Valley's origin and glaciation, the guidebook proved to be a strong bit of advocacy for the preservation of the Yosemite Grant as a national park.

FIGURE 4
Hetch Hetchy Valley with towering Kolana Rock on the right, from a photograph by W. Harris for the Whitney Survey, 1867. *Yosemite Research Library, Hood files.*

In the meantime, in 1867 a new western survey, the "United States Geological Exploration of the Fortieth Parallel," was authorized by Congress to topographically and geologically survey a belt of country over a hundred miles wide and extending from the California state line to the eastern base of the Rocky Mountains. A major part of this terrain encompassed the route of the yet-to-be-completed transcontinental railroad with its enormous economic impact for lands adjacent to the railroad route. With Whitney's endorsement this work was placed under the direction of Clarence King, with long-time colleague James Gardner as his principal topographical assistant, and was carried out with great success.[11] King continued to be interested in glaciation and felt that the 40th Parallel explorations proved that Sierran glaciers were mountain glaciers rather than remainders of a continuous ice sheet as John Muir would have had it. During this period King again climbed and reexamined Mt. Shasta by ascending from a different direction and discovered an actual, active glacier there. He named that glacier for Whitney and published the first description of a "living" glacier in North America (FIG. 5).[12]

FIGURE 5
Clarence King on the Whitney Glacier on Mount Shasta in an 1870 photograph by Carleton Watkins. Watkins was the primary photographer for the Whitney Survey and supplied photographs for Whitney's *Yosemite Book*, 1868. USGS Photo Library.

Perhaps partially in response to King's 40th Parallel Survey, Whitney published an article on "Geographical Surveys" in 1875 that clearly defined the nature of and the need for such surveys. In reviewing the several ongoing surveys in the West, some of which were sponsored by the Department of the Interior and some by the Department of War, he commented on their overlapping nature. He noted that, "Instead of a careful and systematic consolidation of all the United States geographical and geological work in the Far West, under one supervision, in one department, there is just that method employed which leads to bad results and great waste of money."[13]

Whitney's comments were timely in view of ongoing concerns about the duplication of efforts. "Indeed, the matter has already been taken up before a committee of Congress, and a very unpleasant altercation had between the officers and employees of the War Department on one side and of the Interior on the other."[14] The 40th Parallel Survey under the civilian leadership of King was under the Department of the Interior. But it was not until 1879 that such overall consolidation, as recommended by Whitney, occurred when Congress created the U. S. Geological Survey within the Department of the Interior. Clarence King, now with his enhanced reputation, was appointed its first Director, closing the link from Whitney's California Survey through the 40th Parallel Survey to the USGS.

THE UNITED STATES GEOLOGICAL SURVEY

Early Days

In its infancy the USGS began studies in California; one such study included part of the Yosemite region. In 1881 Israel C. Russell (FIG. 6) extended his work on the Mono Lake Basin into the Yosemite High Sierra to study glacial phenomena and provided the first description of the existing glaciers and moraines along this part of the Sierran crest. A highlight of his report is a physiographic map depicting the configuration of the Lyell Glacier at that time, more than 100 years ago (FIG. 7). From that map we can see that the Lyell Glacier today is but a fraction of the size it was then.[15]

FIGURE 6
Israel C. Russell, c. 1900. *USGS Photo Library.*

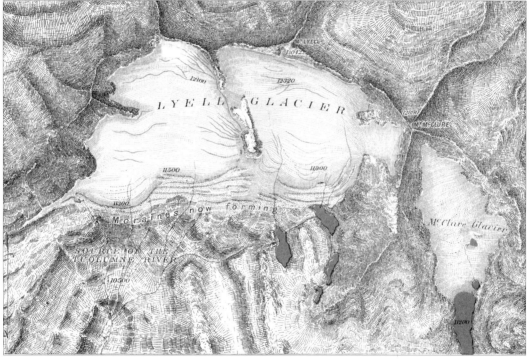

FIGURE 7
Lyell and Maclure Glaciers as they were mapped in 1883 by Willard D. Johnson, a topographer with I. C. Russell's USGS party. Comparison with the present day glaciers indicates their much greater extent at that earlier date. Mount Maclure, named by Whitney for an early American geologist, is here mistakenly labeled "McClure." This error may be the source of occasional confusion in associating the mountain with Lt. N. F. McClure who did not arrive with the cavalry in Yosemite until after 1890. *USGS.*

The year 1886 saw Henry W. Turner and Waldemar Lindgren begin geologic mapping of the California gold belt in the foothills along the west slope of the Sierra Nevada. These studies encompassed a series of 30-minute topographic quadrangle maps that were being geologically mapped moving from north to south along the gold-bearing Mother Lode belt. The work was well along when the Yosemite quadrangle was reached at the southern end of the gold belt. This timing coincided with the establishment of Yosemite National Park in 1890, and Turner (FIG. 8) turned his attention to the study of Yosemite's geology. Turner was the first to record compositional and textural differences among the many varieties of rock in Yosemite and to apply local names to them. Some of those names are still in use today, such as the El Capitan Granite for the variety of rock that makes up that monolith. Turner was also the first to conclude that there had been more than one period of glaciation in Yosemite Valley, establishing the maximum limit of an early glaciation near El Portal, some 10 miles downstream from the valley proper.[16] The lack of economic gold deposits in this region to provide motivation for continued funding support probably doomed Turner's project which was dropped shortly after he started.

Granite domes, hallmarks of Yosemite, are derived by a process of exfoliation, or the peeling off of rock shells. Whitney had recognized that the curvature of the shells was somehow related to the shape of the underlying rock mass. He believed, however, that it was the result of contraction of the granitic material "while cooling or solidifying" rather than being a process related to the present-day topographic surface.[17] The exfoliation process was interpreted by Turner as resulting from cyclic, temperature-related expansion and contraction of the rock mass causing shells eventually to pop loose. His colleague G. K. Gilbert (FIG. 12) emphasized that the rock sheets paralleled the topographic surface, whether on a dome (convex upward) or in a valley (concave upward). Gilbert concluded that expansion toward the exposed surface was due to release of confining pressure in otherwise massive, unfractured granite by sequential removal of overlying material—a process he referred to as "unloading"—an explanation that is generally accepted today.[18]

Over the years a wide discordance developed among various theories for the origin of Yosemite Valley. Some views were extreme, such as Whitney's hypothesis of the catastrophic down-dropping of a block of the Earth's crust and John Muir's belief that Yosemite Valley was gouged out entirely by glaciers. Although less extreme, Turner favored predominant stream erosion while

FIGURE 8
Henry W. Turner, c. 1915.
Geological Society of America.

chief topographer of the USGS Henry Gannett favored predominant glacial erosion. Finally, at the specific request of the Sierra Club, the USGS agreed to undertake a study to resolve the issue, and in 1913 the task fell to François E. Matthes (FIG. 9).

Matthes turned out to be the ideal person to undertake the task. He was a topographer by vocation but a geomorphologist by avocation. His masterful rendering of the topographic map of Yosemite Valley, published in 1907, led him to ponder the valley's landforms, thereby uniquely equipping him to carry out the USGS assignment. His meticulous research resulted in a monumental treatise on the *Geologic History of Yosemite Valley*,[19] which one of the foremost geomorphologists at the time described as a "newborn classic."

Matthes concluded that during at least three glaciations, glaciers advanced down preexisting, V-shaped canyons cut by streams into the western flank of the topographically uprising Sierra Nevada. The glaciers deepened and widened the canyons into U-shapes as they slowly descended, a process supported by modern studies. Thus, both stream and glacial erosion were involved, along with the important role of local rock structure—whether the rock was massive or fractured. In Yosemite Valley itself, however, Matthes greatly underestimated the amount of glacial excavation. Later geophysical studies have revealed that the depth of sediment filling the bedrock basin is as much as 2,000 feet below the present valley floor, a testament to the excavating power of moving ice.[20] But this significant excavation forming Yosemite Valley took place during the earliest glacial episode, which occurred a million years ago. The last glacier to enter the valley, the Tioga, re-excavated only a small amount of the accumulated sediment, leaving a shallow basin containing a lake that extended for some distance upstream in the valley from the moraine dam near El Capitan. Matthes overemphasized the size and depth of this post-glacial lake when he named it "Lake Yosemite" and formalized the lake idea first proposed by Whitney in 1865.[21] Regardless of these problems, Matthes' pioneering studies laid the groundwork for all later geomorphic studies of Yosemite.

FIGURE 9
François E. Matthes in Yosemite, c. 1930. *Photo by Carl P. Russell,* Yosemite Research Library.

One of the more significant of Matthes' contributions is a *Map of Glacial and Postglacial Deposits in Yosemite Valley, Mariposa County, California*.[22] In addition to showing glacial and other surficial deposits and outlining individual glacial moraines, this map depicts his reconstruction of the upper limits of the ice surface of glaciers that occupied the valley. His reconstruction has yet to be improved upon. Matthes' older "El Portal stage" glacier is now correlated with the Sherwin glaciation recognized throughout much of the Sierra Nevada. Although he concluded that his younger "Wisconsin stage" was characterized by two glacial maxima, he did not distinguish between them or map their deposits separately.[23] Thus his "Wisconsin" includes both the Tahoe and the Tioga glaciations of today's terminology. Also included in his volume is a map showing the extent of so-called "Ancient Glaciers" of both his El Portal and Wisconsin glaciations in a broad area reaching east from El Portal to the headwaters of the Tuolumne and Merced Rivers and Illilouette Creek. This map includes the entire area feeding ice to the Yosemite Valley glaciers.[24]

Matthes recognized that "the nature of the Yosemite problem . . . involves two questions: the more specific one of the evolution of the Yosemite Valley, and the more general one of the geologic history of the Sierra Nevada, the mountain range in which the Yosemite lies hewn."[25] To answer these questions, Matthes studied glacial history and the evolution of Yosemite's geomorphology, while his colleague Frank Calkins (FIG. 10) unraveled the bedrock geology. Taking up where Turner left off, Calkins mapped Yosemite's mosaic of discrete bodies of granitic rock by identifying the subtle natural boundaries that separate them. These observations allowed him to establish relative ages of the sequential emplacement of the granitic bodies that together comprise what today we call the Sierra Nevada batholith—the massive granitic backbone of the range. He produced a *Generalized Geologic Map of Part of the Yosemite Region, California*, covering a broad swath stretching from Arch Rock Ranger Station in the west to Tuolumne Meadows in the east.[26] His more detailed *Bedrock Geologic Map of Yosemite Valley, Yosemite National Park, California*, using Matthes' 1907 topographic map as a base, was published posthumously. Although his mapping was done during the period 1913–1916, Calkins was the ultimate perfectionist and never was satisfied that he had everything correct. His mapping has never been improved upon, and finally in 1985 his map was published as a "historical document."[27] It remains the definitive geologic map of Yosemite Valley.

FIGURE 10
Frank C. Calkins, 1907, shortly before beginning his Yosemite studies.
USGS Photo Library.

In summary, Matthes and Calkins were an exceptionally complementary team (FIG. 11). Calkins' mapping of different rock types allowed Matthes to identify the sources for many rock fragments found in the glacial moraines he was tracing. In fact older glacial deposits that were deeply weathered and had lost their original sharp-crested morainal form could only be identified as such by the presence of boulders exotic to their present location. In other words, these erratic boulders were of a different kind of rock than the bedrock upon which they rested, indicating their transport from a distant source to their present site (FIG. 12).

FIGURE 11
François Matthes, Sidney Paige, and Frank Calkins on horseback at their camp at Hog Ranch near present-day Mather, 1916. *USGS Photo Library.*

Later Days

In 1946 under the leadership of Paul Bateman (FIG. 13), the USGS began geologic studies in the Bishop tungsten district of the eastern

FIGURE 12
Large erratic boulder on Moraine Dome. The boulder is composed of Cathedral Peak Granodiorite, readily recognized by the big feldspar crystals projecting from its surfaces. The pedestal consists of a remnant of a shell detached by exfoliation from the body of the dome, which is composed of Half Dome Granodiorite. G. K. Gilbert stands by the boulder. *USGS Photo Library.*

FIGURE 13
Paul Bateman taking a break during geologic mapping in the eastern Sierra Nevada, 1958. *Courtesy of Paul Bateman.*

FIGURE 14
Clyde Wahrhaftig, c. 1975. *Courtesy of Wahrhaftig family.*

Sierra Nevada southeast of Yosemite. Over the next four decades this work evolved from a mineral-resource-oriented project into a broader regional study of the geology across the central Sierra Nevada involving Bateman and more than a dozen colleagues. Armed with a new series of 15-minute topographic base-maps at a scale of 1 inch/mile on which to plot the geology, this project resulted in the preparation of modern, detailed geologic maps covering a large area and eventually including all of Yosemite National Park. At the park's latitude, a 15-minute quadrangle occupies about 240 square miles, and it takes all or part of 15 of them to cover the park and its immediate vicinity.

I was privileged to be part of the Yosemite mapping group that, besides Bateman, included Dallas Peck, who was later to become Director of the USGS. In addition to their geologic mapping, Ron Kistler performed isotopic dating of the granitic rocks, and Frank Dodge studied the mineralogy of those rocks. Clyde Wahrhaftig (FIG. 14), USGS and Professor at University of California, Berkeley, began mapping the Tower Peak quadrangle in northern Yosemite as a personal project in 1955 and was provided USGS support for its completion. In his role as an educator Clyde wrote field guides for the Park Service trail crews and brought them rock samples to leave around the cook tent for their geological edification. In turn he and his students were invited into the trail camp to get a really full meal for a change.

The availability of all the excellent new geologic mapping, supported by the wealth of equally new geophysical, geochemical, and isotopic age data, prompted the USGS in cooperation with the National Park Service to support the compilation of the first modern geologic map completely covering Yosemite National Park. I became the fortunate recipient of that assignment. With his background in glacial geology, Clyde was enlisted to help unravel Yosemite's glacial geology inasmuch as Matthes had worked out the extent of glaciers for only a limited area. Bateman had been independently working on the correlation of the various granitic bodies that make up this part of the Sierra, and his work formed the basis for much of the explanatory material for the new map.[28]

Although at this time several of the pertinent geologic quadrangle maps within Yosemite had yet to be published, we had access to all of the necessary information. Putting it together was not a simple task because each of the individual quadrangles had been mapped by different groups at different times and in random, checker-board pattern within the park rather than in a sequential geographic order of adjacent quadrangles. Geologic mapping is, at least in part, an interpretive process, especially at the scale involved. It is also made more complicated at mid- to lower-elevations by forest cover. As a result, numerous problems arose regarding different interpretations in adjacent quadrangles, and I spent two summers in the park trying to resolve them. Clyde also had his hands full since most of the quadrangle mappers had concentrated on the bedrock formations and showed glacial deposits only where they completely obscured the bedrock.

The resulting product of our efforts was the *Geologic Map of Yosemite National Park and Vicinity*, which synthesizes the geology of more than 2,000 square miles of the central Sierra Nevada centered on Yosemite.[29] Production of this map involved reducing and somewhat generalizing or simplifying the geology of the 15-minute quadrangles from a scale of 1 inch/mile to a smaller scale of ½ inch/mile in order to get everything on one sheet. This multi-color map depicts the park's geology with 70 individual geologic map units including surficial deposits and volcanic, sedimentary and metamorphic rocks. Granitic rocks of many varieties dominate the Yosemite landscape, and together they are part of the Sierra Nevada batholith, a composite body of granitic rock stretching the length of the range. Such granitic rocks make up more than 80% of the map area. The map explanation provides a brief description of each geologic unit, and a brief summary highlights some key features of the park's geology.

Although we had learned a great deal resulting in significant advances in geologic concepts since Matthes' time, there was little modern material on Yosemite geology suitable for developing a well-balanced geologic interpretive program for the park visitor. Filling this void was a major goal of the Yosemite project, and the vehicle settled on was a book titled, appropriately enough, *The Geologic Story of Yosemite National Park*.[30] This book, which I was privileged to write, synthesizes the geology of Yosemite with a minimum of technical jargon and is the first to treat all major aspects of the park's geology. It is also the first to apply modern plate-tectonic concepts to the

FIGURE 15

Tau Rho Alpha sketching in Yosemite Valley, 1985. *Courtesy of Tau Rho Alpha.*

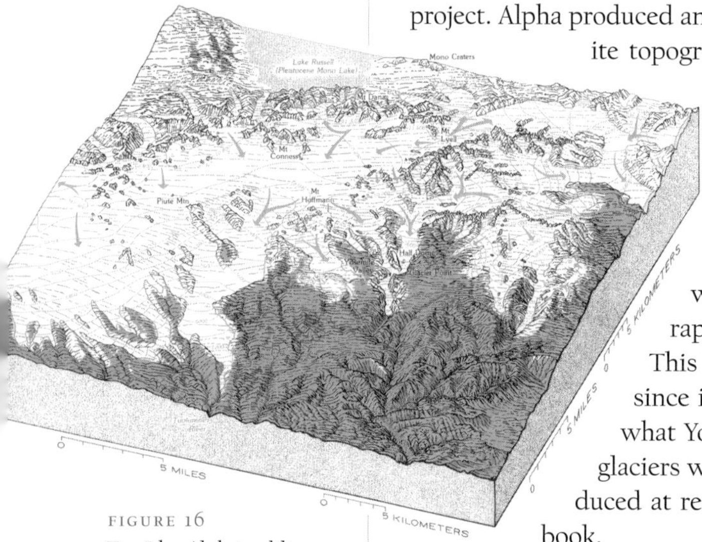

FIGURE 16

Tau Rho Alpha's oblique map of icefield and valley glaciers showing maximum extent during the Tioga glaciation which peaked about 20,000 to 15,000 years ago, the last major glaciation in the Sierra Nevada, from Huber, 1987.

origin of Yosemite's granitic terrane. The book describes the rock formations and their origins and fits them into the context of a geologic history.

My long career of geologic studies in the Sierra provided me with the background to put Yosemite's geology into a regional perspective. But writing my Yosemite book was not easy. We professionals traditionally write for our peers who are familiar with technical jargon. But the jargon must be avoided to make explanations clear for the general public. Hitting this middle ground was not easy. Fortunately I had the help of talented editors and illustrators within the USGS to help put the book together. Especially helpful was cartographer Tau Rho Alpha (FIG. 15), who produced the fold-out view from Mt. Hoffmann and other illustrations. The *Geologic Map of Yosemite National Park and Vicinity* is a good companion to the book because the book itself contains only a much generalized, small-scale geologic map as well as a generalized version of Calkins' geologic map of Yosemite Valley.

Some additional products of note emanated from our Yosemite project. Alpha produced an oblique block diagram depicting Yosemite topography,[31] which was also used at reduced scale as FIGURE 6 in my Yosemite book to locate Mt. Hoffmann. Clyde and I joined with Tau to produce a second oblique block diagram showing the maximum extent of glaciers of the last major glaciation (20,000 years ago), which was superimposed on the topography of the previous Yosemite diagram.[32] This is a favorite of the park interpretive staff since it gives them a stunning, graphic view of what Yosemite might have looked like when the glaciers were here (FIG. 16). This map is also reproduced at reduced scale as FIGURE 67 in my Yosemite book.

A final word. During my personal sojourn in the park there developed a mutually productive relationship with the Park Service scientific and interpretive staff, an essential ingredient to the success of our project. This was truly a cooperative effort between the U. S. Geological Survey and the National Park Service without whose logistical support the necessary field work would have been impossible. Although we can claim something close to state-of-the-art for our geologic products, we also know that the final word will never be written. Theories evolve, new tools will be invented, and new field discoveries, heretofore overlooked, will be made. We know that those who follow us will continue to advance these new frontiers as did those of us who followed in the footsteps of Josiah Whitney and Clarence King.

NOTES

1. J. D. Whitney, 1865, *Geology, Volume 1, Report of Progress and Synopsis of the Field-Work from 1860 to 1864* (Geological Survey of California, by Authority of the State Legislature).
2. Whitney, 1865, p. 428-429.
3. Whitney, 1865, p. 423.
4. Whitney, 1865, p. 421.
5. Hank Johnston, 1995, *The Yosemite Grant, 1864-1906: A Pictorial History* (Yosemite National Park, CA, Yosemite Association). Appendix B contains the complete text of the congressional act authorizing the Yosemite Grant.
6. J. D. Whitney, 1868, *The Yosemite Book* (Geological Survey of California, New York, Julius Bien). Regarding the photographs, at that time photolithography was not yet available and only contact prints the exact size of the negative could be made.
7. J. D. Whitney, 1869, *The Yosemite Guide-Book* (Cambridge, MA, Harvard University Press) and later editions.
8. J. D. Whitney, 1880, *The Climatic Changes of Later Geologic Times: A Discussion Based on Observations Made in the Cordilleras of North America* (Cambridge, MA, Harvard University Press). See also "From V to U—Glaciation and Valley Sculpture," this volume.
9. Whitney, 1869, p. 111-112.
10. Whitney, 1869, p. 112. See also "A Tale of Two Valleys," this volume.
11. J. G. Moore, 2006, *King of the 40th Parallel, Discovery in the American West* (Stanford, CA, Stanford University Press).
12. Clarence King, 1871, *Active Glaciers within the United States* (Atlantic Monthly, v. 27, no. 161), p. 371-377. Another version, "Shasta," became Chapter 11 in his *Mountaineering in the Sierra Nevada* (1872).
13. J. D. Whitney, 1875, *Geographical and Geological Surveys. I. Geographical* (North American Review, v. 121, no. 248), p. 83-84.
14. Whitney, 1875, p. 83.
15. I. C. Russell, 1889, *Quaternary History of Mono Valley, California* (U. S. Geological Survey Eighth Annual Report, p. 261-394, reprinted by Artemesia Press, Lee Vining, CA, 1984), p. 261-394.
16. H. W. Turner, 1900, *The Pleistocene Geology of the South-Central Sierra Nevada, with Especial Reference to the Origin of the Yosemite Valley* (California Academy of Sciences Proceedings, 3rd series, Geology, v. 1), p. 261-321.
17. Whitney, 1865, p. 371-372.
18. G. K. Gilbert, 1904, *Domes and Dome Structure of the High Sierra* (Geological Society of America Bulletin, v. 15), p. 29-36.

19. F. E. Matthes, 1930, *Geologic History of the Yosemite Valley* (U. S. Geological Survey Professional Paper 160).
20. Beno Gutenberg, J. P. Buwalda, and R. P. Sharp, 1957, *Seismic Explorations on the Floor of Yosemite Valley, California* (Geological Society of America Bulletin, v. 67), p. 1051–1078. See also "From V to U—Glaciation and Valley Sculpture" and "How Deep Is the Valley?" in this volume.
21. Matthes, 1930, p. 103.
22. Matthes, 1930, Plate 29.
23. Matthes, 1930, p. 60–61.
24. Matthes, 1930, Plate 39.
25. F. E. Matthes, 1914, *Studying the Yosemite Problem* (Sierra Club Bulletin, v. 9, no. 3), p. 138.
26. F. C. Calkins, 1930, *Generalized Geologic Map of Part of the Yosemite Region, California* (Matthes, 1930, Plate 51).
27. F. C. Calkins, 1985, *Bedrock Geologic Map of Yosemite Valley, Yosemite National Park, California*, with accompanying pamphlet by N. K. Huber and J. A. Roller (U. S. Geological Survey Miscellaneous Investigations Series Map I-1639).
28. P. C. Bateman, 1992, *Plutonism in the Central Part of the Sierra Nevada Batholith, California* (U. S. Geological Survey Professional Paper 1483).
29. N. K. Huber, P. C. Bateman, and Clyde Wahrhaftig, 1989, *Geologic Map of Yosemite National Park and Vicinity, California* (U. S. Geological Survey Miscellaneous Investigations Series Map I-1874).
30. N. K. Huber, 1987, *The Geologic Story of Yosemite National Park* (U. S. Geological Survey Bulletin 1595, reprinted by Yosemite Association, 1989).
31. T. R. Alpha, Clyde Wahrhaftig, and N. K. Huber, 1986, *Oblique Map of Yosemite National Park, Central Sierra Nevada, California* (U. S. Geological Survey Miscellaneous Investigations Series Map I-1776), was used in reduced version as Figure 14 in Huber, 1987.
32. T. R. Alpha, Clyde Wahrhaftig, and N. K. Huber, 1987, *Oblique Map Showing Maximum Extent of 20,000-Year-Old (Tioga) Glaciers, Yosemite National Park, Central Sierra Nevada, California* (U. S. Geological Survey Miscellaneous Investigations Series Map I-1885), was used in reduced version as Figure 67 in Huber, 1987.

A Glacial Footnote

BY N. KING HUBER AND JIM SNYDER

FOR MORE THAN ONE HUNDRED AND FORTY YEARS, EVIDENCE HAS BEEN gathered about the extent and role of glaciers in Yosemite National Park. Evidence for the presence of past glaciers in Yosemite and elsewhere in the Sierra Nevada was noted quite early and became public through the interactions of three remarkable men.

JOSIAH WHITNEY AND CLARENCE KING

California State Geologist Josiah D. Whitney (FIG. 1) and members of his Geological Survey of California observed "abundant traces of glacier action" with "immense flows" of ice commonly of "immense thickness" in many places between the headwaters of the Kern and the Mokelumne Rivers.[1] These and similar statements in Whitney's *Geology, Volume I* constitute the first comprehensive description of glacial evidence in the Sierra. While news of glacial evidence was accepted by many, the question of the extent of glaciation remained open, and debate focused on Yosemite Valley. In this same report, which became available in early 1866,[2] Whitney referred to two members of his survey crew when he stated: "In the course of the explorations of Messrs. [Clarence] King and [James] Gardner, they obtained ample evidence of the former existence of a glacier in the Yosemite Valley, and the cañons of all of the streams entering it are also beautifully polished and grooved by glacial action. It does not appear, however, that the mass of ice ever filled the Yosemite to the upper edge of the cliffs: but

FIGURE 1
Josiah Whitney, 1863.
Courtesy of the Bancroft Library (CA Geological Survey—POR 1), University of California, Berkeley.

Mr. King thinks it must have been at least a thousand feet thick. He also traces out four ridges in the valley which he considers to be, without a doubt, ancient [glacial] moraines."[3]

Whitney, however, did not entirely trust the observations of his young assistants and could not accept a glacial explanation of Yosemite Valley's origin. In part, he became jealous of King's success in landing a major federal survey of his own, but there were also legitimate questions about some of King's observations and mapping for the California Survey. Differences between the more obviously glaciated Hetch Hetchy Valley and the ragged, weathered cliffs of Yosemite Valley, for example, seemed to argue forcefully against Yosemite as a glaciated valley.

Therefore, in his *Yosemite Book* published in 1868, Whitney repudiated his earlier statements regarding King's observations: "We have obtained no evidence that such [glaciation in Yosemite Valley] was the case. The statement to that effect in the 'Geology of California,' Vol. I, is an error, although it is certain that the masses of ice approached very near to the edge of the Valley."[4] In 1880 Whitney reinforced his rebuttal by writing: "The walls of the Yosemite on each side were carefully examined by the writer without his having been able to find on them any sign of smoothed, striated, or polished surfaces which could be unhesitatingly set down as the work of ice."[5]

In the meantime, in 1870, King (FIG. 2) discovered, described, and had Carleton Watkins photograph the first "living" glacier found on the west coast, and named that glacier on Mt. Shasta for Josiah Whitney. It is obvious that King's respect for his former chief had not diminished. The first discovery of an active glacier in the Yosemite region was made the following year by John Muir on "Black Mountain" in the Clark Range at the headwaters of Illilouette Creek.

In his *Mountaineering in the Sierra Nevada* (1872), King resurrected his earlier observations about Yosemite Valley, which had been first cited and then rejected by Whitney, and wrote: "The markings upon the glacial cliff above Hutchings' house had convinced me that a glacier no less than a thousand feet deep had flowed through the valley, occupying the entire bottom."[6]

Despite their disagreement over glaciers in Yosemite Valley, King and Whitney shared information for King's *Systematic Geology* (1878). In that work King agreed with Whitney that glaciation in the Sierra had not been a range-covering ice cap, but rather a sequence of valley glaciers. In this, King defended Whitney's position against

FIGURE 2

Clarence King, c. 1865. *Galen Clark album*, Yosemite Museum.

John Muir, who, despite his interesting local observations on glacial action, argued for a very extensive ice cap over the entire Sierra.

FREDERICK LAW OLMSTED

Even though glacial evidence at lower elevations seemed speculative in late 1865 when Whitney's *Geology* was published, information about glaciation in Yosemite Valley had crept out even earlier. In the summer of 1865, prior to the publication of Whitney's *Geology*, no less a person than renowned landscape architect Frederick Law Olmsted (FIG. 3), who designed New York's Central Park, included in a description of Yosemite Valley this line: "At certain points the walls of rock are ploughed in polished horizontal furrows, at others moraines of boulders and pebbles are found; both evincing the terrific force with which in past ages of the earth's history a glacier has moved down the chasm from among the adjoining peaks of the Sierras."[7] How could this have come about?

It turns out that Olmsted at that time was superintendent of John C. Frémont's Mariposa Estate, a gold-mining property then owned by New York interests and located just west of Yosemite. Production at the mines had slumped, and Olmsted, with his lack of experience in mining, looked for expert counsel and turned to Whitney for help. Whitney saw a chance to study an important gold-vein region and assigned two of his assistants to the task: William Ashburner, to pursue problems of mining and engineering, and Clarence King, to conduct a general survey of the geology, with the hope of finding fossils to date the rocks.

The two men left for the estate in late November 1863, with King looking forward to a reunion with Olmsted, whom he had known from his days back east in Hartford.[8] In addition to the Olmsteds, King also got to know the Olmsted family governess, Harriet Errington, who became interested in geology and fossils, one of which was subsequently named in her honor. Errington accompanied King, sometimes with Olmsted, on several day-long exploring trips on the Mariposa Estate, in the Mariposa Grove, and in Yosemite Valley. Their discoveries and observations were matters of family discussion.[9] In addition to Olmsted and Errington, the affable King conversed with other locals, such as the Hutchings family who ran a hotel in the valley and Galen Clark who ran an establishment at the site of present-day Wawona.

The Congressional Act granting Yosemite Valley and the Mariposa Grove of Big Trees to the State of California was signed by President Lincoln in June 1864. A Board of Commissioners, which

FIGURE 3
Frederick Law Olmsted, c. 1865. *Galen Clark album, Yosemite Museum.*

included the Governor and eight others, was created to manage the Yosemite Grant. Olmsted fortunately was appointed to the commission and became its functional head.[10]

Even before his formal appointment, Olmsted personally hired Clarence King and James Gardner, both of the California Survey, to survey and establish the boundaries of the grant, which they accomplished in the fall of 1864. It was during this survey that King made the observations that led him to conclude that the valley once hosted a glacier at least "a thousand feet thick," as quoted in both Whitney's and King's publications. Thus conversations with King were most likely the basis for Olmsted's statement in his 1865 report, since nothing on glaciation had yet been published, and Olmsted would not have made such observations on his own.

For several reasons, Olmsted's brief statement regarding glaciation in Yosemite Valley never received the attention it deserved. First, his statement was made before the debate about such glaciation really developed. Second, his comment was almost an offhand remark, a brief part of a long introduction to his report containing farsighted recommendations to the California legislature for management of the Yosemite Grant. His report was first presented orally to commissioners William Ashburner, Galen Clark, George Coulter, and Alexander Deering while camped in Yosemite Valley in August 1865. The other members of the commission, including Whitney and Governor Low, were not present at this meeting.

Olmsted's report was adopted by the commission, but that fall, after Olmsted returned to New York, three members of the commission, including Whitney and Ashburner, forwarded the report to the Governor. They recommended not sending the report on to the Legislature, in part because of its request for a substantial sum to build a road to Yosemite, which they felt would not be expedient while the State was having severe budget problems. The Governor apparently agreed, and Olmsted's proposals were neither published nor funded.[11] The report continued to be circulated in manuscript form into the 1890s, but the long description of Yosemite Valley, including Olmsted's brief mention of glaciation there, did appear in an 1868 letter of Olmsted's to the *New York Evening Post*.[12] Olmsted's full report did not reach a general audience until its eventual publication in 1952 by Olmsted's biographer, Laura Wood Roper.[13]

In the end, despite his rejection of a glacial theory of Yosemite Valley's origin, Whitney inadvertently had a hand in making his *Geology* the first published report to present evidence for glaciation in Yosemite Valley.

NOTES

1. Josiah D. Whitney, 1865, *Geology, Volume I, Report of Progress and Synopsis of Field-Work, From 1860 to 1864* (Philadelphia, Sherman & Co., for the Legislature of California), pages 372, 425, and 447, for example.
2. Robert H. Block, 1982, *The Whitney Survey of California: A Study of Environmental Science and Exploration* (Ph.D. dissertation, UCLA, 1982). Page 367 cites a January 19, 1866, letter from Whitney to his brother saying that the first two volumes of the Survey, *Paleontology* and *Geology Volume I*, were distributed in San Francisco beginning January 15, 1866. Whitney's preface to *Geology Volume I*, written prior to publication, is dated November 1, 1865.
3. Whitney, 1865, *Geology*, p. 422.
4. Josiah D. Whitney, 1868, *The Yosemite Book* (New York, Julius Bien for the Legislature of the State of California), p. 100; *The Yosemite Guide-Book* (Cambridge, University Press, 1869), p. 112.
5. Josiah D. Whitney, 1880, *The Climatic Changes of Later Geological Times: A Discussion Based on Observations Made in the Cordilleras of North America* (Cambridge, Harvard University Press), p. 47.
6. Clarence King, 1872, *Mountaineering in the Sierra Nevada* (Lincoln, University of Nebraska Press, 1970; orig. pub. New York: James R. Osgood & Co., 1872), p. 152.
7. Frederick Law Olmsted, 1865, *The Yosemite Valley and the Mariposa Big Trees: A Preliminary Report*, edited by Victoria Post Ranney (Yosemite Association, 1995), p. 4.
8. Thurman Wilkins, 1988, *Clarence King, A Biography* (Albuquerque, University of New Mexico Press, Revised Edition), p. 57.
9. Frederick Law Olmsted, Jr., compiler, 1952, "Harriet Errington's Letters and Journal from California 1864-65" (typescript in Yosemite Research Library, 1952).
10. Hank Johnston, 1995, *The Yosemite Grant, 1864-1906* (Yosemite Association), p. 62.
11. Johnston, p. 67.
12. Frederick Law Olmsted, 1868, "The Yosemite Valley," *New York Evening Post*, June 18, 1868, p. 29–30. See also Victoria Post Ranney, 1990, *The Papers of Frederick Law Olmsted, Vol. 5: The California Frontier 1863–1865* (Baltimore, The Johns Hopkins University Press), p. 491, 511, 514.
13. Laura Wood Roper, 1952, "The Yosemite Valley and the Mariposa Big Trees, A Preliminary Report (1865) by Frederick Law Olmsted," *Landscape Architecture*, vol. 43, no. 1 (October 1952), p. 12–25. See also Laura Wood Roper, 1973, *FLO, A Biography of Frederick Law Olmsted* (Baltimore, The Johns Hopkins University Press), p. 282-288.

Tracking the Fire

Evolution of the Tuolumne River

BY N. KING HUBER

HETCH HETCHY VALLEY AND YOSEMITE VALLEY HAVE BEEN COMPARED BY many, including the eloquent John Muir. Muir tended to emphasize their similarities. In fact, he used "Yosemite" as a generic term and stated that "the Tuolumne Yosemite is a wonderfully exact counterpart of the Merced Yosemite." For each he emphasized their bold cliffs, waterfalls, and flat valley floors. In spite of this, the two valleys are quite different in one important way. The Merced River drops abruptly to Yosemite Valley at Nevada and Vernal Falls, as does Tenaya Creek at Pywiack Cascades. In contrast, the Tuolumne River drops more gradually over a longer distance into Hetch Hetchy Valley through the Grand Canyon of the Tuolumne, below a series of cascades west of Tuolumne Meadows. The difference can be attributed, in part, to differences in the glacial histories of the two valleys.

During each major glaciation, including the last whose maximum was probably only 15,000–20,000 years ago, the Tuolumne canyon was filled to the brim with ice at least as far west as Mather, some 6 miles beyond Hetch Hetchy (FIG. 1). Thus, Hetch Hetchy has been glacially scoured "recently."

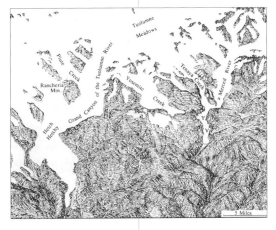

FIGURE 1

Diagram showing Tioga-age glaciers in the Tuolumne and Merced drainages. YV, Yosemite Valley.

Yosemite Valley, however, has not been completely filled with ice for at least one million years, the minimum age of the Sherwin glaciation (probably equivalent to Matthes' El Portal glaciation[1]). The major excavation of Yosemite Valley, including the bedrock basin beneath the valley floor, had to have been accomplished by that time. Since then, the upper reaches of Yosemite Valley cliffs have been shaped by spalling rather than by glacial scour, leaving pinnacles that could not survive a valley-full glaciation and forming the recessed alcoves into which waterfalls such as Bridalveil now leap. For this reason Hetch Hetchy Valley is in many ways a "fresher-looking" glaciated valley than Yosemite Valley, which long has been considered a classic glacially-carved valley.

These differences probably helped to stoke the Muir-Whitney confrontation. Yosemite Valley, in spite of its profound glacial modification, is not a good place for visual evidence of glaciation. The Tioga (latest) glacier reached only as far as Bridalveil Meadow, where it deposited a relatively inconspicuous terminal moraine. No lateral moraines could survive, because they were against the precipitous walls and were quickly eroded. Easily accessible outcrops with glacial polish and striations are scarce. This paucity of direct evidence is probably the basis for Whitney's stand against the glacial origin of Yosemite Valley, noting as he did the glaciation of Hetch Hetchy Valley, where he described glacial polish at least 800 feet above the valley floor and lateral moraines 1,200 feet above the floor on the upland beyond the valley rim.[2]

FIGURE 2

Ten-million-year-old (Miocene) volcanic and fluvial deposits in the Tuolumne drainage basin and reconstruction of the ancient stream channel. Inset shows longitudinal profiles of modern and reconstructed Miocene channels projected into a southwest-northeast line. Arrows indicate control points for reconstruction.

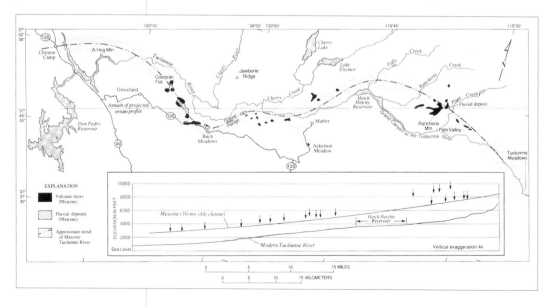

A recent study of the late Cenozoic evolution of the Tuolumne River[3] reveals more details about the geomorphic development of the Sierra Nevada itself. The uplift and westward tilt of the range was underway by at least 15 million years ago,[4] and the steepening stream gradients accelerated canyon incision. During this time, volcanic eruptions in the Sierra north of Yosemite buried the drainages of west-flowing streams with volcanic debris, and the rivers were forced to cut new courses to the Central Valley.

About 10 million years ago, some of this volcanic material, chiefly volcanic mudflows, flowed south past the present drainage divide from the Stanislaus into the Tuolumne drainage basin. This material entered the ancient Tuolumne channel near Rancheria Mountain and flowed westward within that channel.

Most of this material has since been removed by erosion, but remnants permit us to document the location of the ancient channel and reconstruct its subsequent history (FIG. 2). The river was forced to shift laterally southward around the volcanic "dam" near Rancheria Mountain and cut a new channel next to the volcanic infilling. Because the volcanic material did not overflow the valley of the ancient Tuolumne, as it did with rivers further north, the present channel follows the ancient one fairly closely.

FIGURE 3
Ancient channel of the Tuolumne River exposed along the western bank of Piute Creek at Rancheria Mountain. The river flowed westward away from the viewer into the V-shaped notch cut into granite. Stream gravel in the channel was later buried beneath volcanic mudflows. This ancient channel was first described by H. W. Turner who took this photograph about 1900. *USGS Photo Library.*

One of the most convincing displays of the ancient channel is on Rancheria Mountain west of Piute Creek (FIG. 3). Here volcanic mudflows in the ancient channel overlie stream gravel that contains pebbles of metamorphic rock derived from as far east as Mount Dana. These pebbles indicate that the Tuolumne River was draining approximately the same headwaters 10 million years ago that it does today.

Ten million years ago, an ancestral range of hills occupied the present site of the Sierran crest, and, although of relatively moderate relief, it was a barrier to westward drainage even before late Cenozoic uplift. At that time, the San Joaquin River was apparently the only river flowing westward across the part of the range that lies south of Sonora Pass. The drainage of this ancient San Joaquin River included the area now occupied by the Mono basin. The ancient Tuolumne River evidently never extended east of this range of hills. Neither the reconstructed Tuolumne channel nor its deposits in the Sierra nor its alluvial fans in the Central Valley indicate or even suggest a source east of the present range.

Although the Tuolumne River apparently never headed east of the present range, one of its forks may have headed east of Tioga Pass. The trough containing Tioga Pass trends north-south and nearly aligns with the valley of the upper part of Lee Vining Creek and Saddlebag Lake (FIG. 4). A profile down this upper creek and through the trough of Tioga Pass dips only 500 feet below the pass, and the pass itself contains an unknown thickness of glacial till. If upper Lee Vining Creek once drained south through Tioga Pass, it was subsequently captured by the main trunk of Lee Vining Creek during its headward incision, possibly aided by a glacier or glaciers flowing eastward over a saddle in the former crest, and the route over Tioga Pass would have been abandoned. Before this capture, the Sierran drainage divide would have been a few miles east of the present one between Mount Dana and Excelsior Mountain, following a belt of resistant metamorphic rocks over Dana Plateau, Tioga Peak, and Tioga Crest (FIG. 4). The present abrupt drop of Lee Vining Creek below Ellery Lake takes place at the eastern edge of this metamorphic belt.

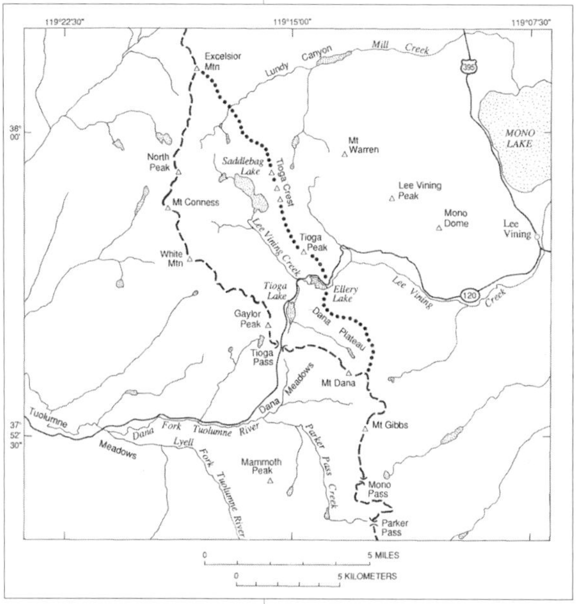

FIGURE 4

Tioga Pass area showing postulated earlier drainage divide (dotted line). Before capture by Lee Vining Creek, the basin containing Saddlebag Lake might have drained south through Tioga Pass to the Tuolumne River. Present drainage divide shown as dashed line.

The San Joaquin River was a sufficiently large river to maintain its course across the rising Sierra, until it was cut off by volcanic activity about 3 million years ago. As a result its present channel is deeply incised right up to the Sierran divide. In contrast, the Tuolumne River is deeply incised eastward only to a point several miles west of Tuolumne Meadows. From there eastward to the Sierra crest the river meanders through upland meadows and up the broad low-gradient Lyell and Dana Forks. With no trans-Sierra drainage, uplift-induced incision of the Tuolumne channel proceeded headward from the Central Valley: major incision has not yet reached Tuolumne Meadows.

In the preceding discussion of the evolution of the Tuolumne River system, emphasis was placed on the main trunk of the river

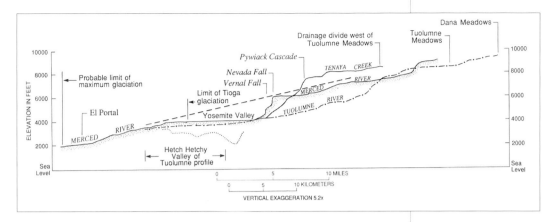

FIGURE 5
Longitudinal profiles of Merced River and tributary Tenaya Creek, with Tuolumne River for comparison. All profiles follow stream courses, but ignore minor meanders. The Tuolumne plot is superimposed to intersect at the 4,000-ft elevation, the elevation of the Merced-Tenaya junction at the head of Yosemite Valley. The dotted line indicates the bedrock basin in Yosemite Valley, as interpreted from seismic data.[7] The dashed line indicates the inferred average trend of the pre-glacial Merced River without the Yosemite Valley basin, although some excavation below the line probably resulted from stream erosion prior to glaciation.

that heads in its Dana and Lyell Forks—an area critical to Schaffer's recent postulate of past trans-Sierra drainage at this location.[5, 6] Schaffer proposed that a Tenaya "River," existing since early in the Miocene, originated from lands east of today's range and flowed through the site of today's Tioga Pass, Tuolumne Meadows, Tenaya Lake, Tenaya Canyon and Yosemite Valley. He further suggested that the headwaters of this Tenaya River were subsequently captured by the Tuolumne River at Tuolumne Meadows "about one-half to one-quarter million years ago." If the history of the Tuolumne River that I have sketched is correct, then there never was a trans-Sierra drainage through Tioga Pass, and at 10 million years ago the Tuolumne River was already draining the Mount Dana area.

Comparison of present-day stream profiles argues further against the concept of a Tenaya River. The part of the Merced drainage basin (including Tenaya Creek) that drains into Yosemite Valley is about one third the size of that part of the Tuolumne basin that drains into Hetch Hetchy, and the distance to its drainage divide and its discharge are proportionally smaller. Thus the Merced would be expected to develop a less mature (steeper average gradient) stream profile than did the Tuolumne over the same period of time and before modified by glacial erosion. That that is the case is illustrated where parts of the longitudinal profiles for the present-day Merced River (and tributary Tenaya Creek) and the Tuolumne River are compared (FIG. 5). The inferred average profile of the pre-glacial Merced River is also indicated.

The most obvious anomaly on the Merced profile is the "glacial staircase" formed by Nevada and Vernal Falls just east of Yosemite

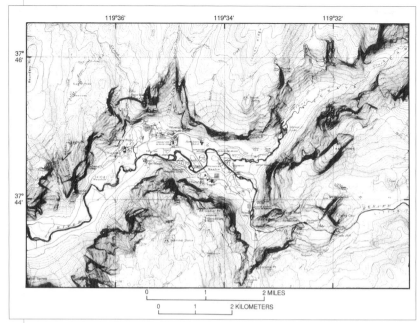

FIGURE 6

Joint-sets control the orientation of physiographic features in Yosemite Valley. Shown here is part of the topographic map of the valley by François Matthes (1907), whose treatment emphasizes lineal and planar features more graphically than does the modern version. USGS.

Valley. There is little doubt that the present form of these steps resulted from differential glacial quarrying of granite that was highly fractured along major joint systems that were downstream from areas of more resistant granite, in the manner envisioned by François Matthes. Less obvious, because it is concealed, is the glacially excavated bedrock basin beneath Yosemite Valley (FIG. 5), which to some degree is an extension of the staircase.

A northeast-trending joint set dominates the Yosemite Valley region (FIG. 6). It is not only strikingly displayed at such features as Nevada Fall and the elongate Half Dome, but also provides directional control for segments of Tenaya Canyon, Yosemite Valley, the Merced Gorge (downstream from Yosemite Valley), and several tributary creeks. A northwest-trending set, displayed at Vernal Fall, also provides control for segments of Yosemite Valley. The preglacial channel of the Merced presumably was influenced by these joint sets, but flowing water does not have the excavating power of glacial ice and its profile would have only relatively minor perturbations reflecting differences in stream erodability of the bedrock.

Tenaya Creek drops abruptly down Pywiack Cascade into Tenaya Canyon before joining the Merced River at the head of Yosemite Valley. Matthes contrasted the narrow, deep Tenaya Canyon with the wider but shallower Little Yosemite Valley on the Merced River just above Nevada Fall. He attributed the geometry of Tenaya Canyon to glacial excavation of a zone of fractured rock along a major joint set; the lower, deepest section of Tenaya Canyon is parallel with the prominent northeast-trending joint set. This interpretation has been challenged by Schaffer who downplays glacial excavation relative to pre-glacial stream incision of Yosemite Valley and Tenaya Canyon.[5, 6] He does not believe that "such a small stream, the Merced River," could accomplish the amount of pre-glacial incision in Yosemite Valley that his supposition requires and, therefore, proposes his trans-Sierra Tenaya River to have existed until less than a million years ago to excavate the valley.

The stream profile for Tenaya Creek, however, has a low gradient for only about one and a half miles above its junction with the Merced River. It then climbs rapidly to rise above the profile of the Merced River only four miles above the junction, and in a little more than five miles is 1,400 feet higher in elevation than the profile of the Merced. This, and other evidence presented above, suggests that throughout late Cenozoic time, Tenaya Creek has been tributary to the larger Merced River and has not had a significantly larger drainage basin than it has now. Matthes' explanation for the depth and orientation of Tenaya Canyon, through the guidance of glacial excavation along joints, still seems reasonable.

I conclude that the geometry of the longitudinal stream profile of the Tuolumne River from Hetch Hetchy eastward reflects pre-glacial stream erosion, moderately modified by glacial erosion, mainly in the reach of the cascades below Tuolumne Meadows. In contrast, the very anomalous profiles of the Merced River through Yosemite Valley and eastward, and that of its tributary Tenaya Creek, are almost entirely the result of glacial excavation.

(Revised from *Yosemite*, v. 52, n. 1 [Winter 1990])

NOTES

1. François E. Matthes, 1930, *Geologic History of the Yosemite Valley*, U. S. Geological Survey Professional Paper 160.
2. Josiah D. Whitney, 1865, *Geology, Volume I*, Geological Survey of California.
3. N. King Huber, 1980, "The Late Cenozoic Evolution of the Tuolumne River, Central Sierra Nevada, California," *Geological Society of America Bulletin*, v. 102, p. 102–115.
4. N. King Huber, 1981, *Amount and Timing of Late Cenozoic Uplift and Tilt of the Central Sierra Nevada—Evidence from the Upper San Joaquin River Basin*, U. S. Geological Survey Prof. Paper 1197, 28 p.

5. Jeffrey P. Schaffer, 1986, "A Geologic History of Yosemite Valley," printed on the reverse side of *Topographic Map of Yosemite Valley* (Berkeley, CA, Wilderness Press).
6. Jeffrey P. Schaffer, 1987, "A New Look at the Origin of Yosemite Valley," Yosemite Association, *Yosemite*, v. 49, no. 3, p. 6-8.
7. Beno Gutenberg, J. P. Buwalda, and R. P. Sharp, 1956, "Seismic Explorations on the Floor of Yosemite Valley," *Geological Society of America Bulletin*, v. 67, p. 1051-1078.

James Mason Hutchings and the Devils Postpile

BY N. KING HUBER AND JIM SNYDER

The "Devils Postpile" in the eastern Sierra Nevada of California is an erosional remnant of a lava flow that erupted within the canyon of the Middle Fork of the San Joaquin River sometime between 50,000 and 100,000 years ago. About 20,000 years ago, a glacier moving down the river valley removed the upper part of the lava flow to expose its interior and display a magnificent example of columns formed by shrinkage during the cooling and solidification of the once-molten lava flow (FIG. 1).

The Devils Postpile was originally included within Yosemite National Park when the park was established in 1890. The Postpile was deleted from the park when the park boundaries were realigned in 1905, but was returned to the National Park System as a National Monument in 1911. But in 1875, 15 years before establishment of the park, James Mason Hutchings chanced upon the Postpile and made some remarkable observations about it—observations that have never before appeared in print.

FIGURE 1
The Devils Postpile. *Photo by Gerhard Schumacher, USGS Photo Library.*

Hutchings is best known for his association with Yosemite Valley, which began very early. He led the first "tourist" party to the valley in 1855, only four years after non-Indians first entered the valley. His association with the valley extended over many years, and included the operation of a hotel there. He is also widely known for his many publications, including *Scenes of Wonder and Curiosity in California* first published in 1860, a major part of which was "A Tourist's Guide to the Yo-Semite Valley."

In true entrepreneurial spirit, Hutchings organized an expedition in 1875 to journey from Yosemite Valley to Mount Whitney in order to take the first photographs from its summit. Mount Whitney, the highest peak in the lower 48 states at 14,494 feet, was first climbed only two years earlier. Hutchings apparently intended to publish a book using these and other photographs taken along the expedition's route. Although they were never published, Hutchings used the photographs in a number of lantern-slide lectures in the 1870s and 1880s.

The photographer on the expedition was W. E. James, about whom little is known. He was one of a number of photographers that Hutchings used to illustrate his publications and lectures. From San Francisco, where he apparently met Hutchings, James moved to Santa Cruz in the 1880s, and then dropped from sight. Dr. Albert Kellogg, physician, botanist, and namesake of *Quercus kelloggii*, the California black oak, was also a member of the expedition to Mount Whitney.

Describing the expedition's eastbound traverse across the Sierra Nevada from Yosemite Valley to Long Valley, Hutchings' diary indicates that they camped near the Devils Postpile and took a photograph of it. In the early 1930s Hutchings' daughter "Cosie" transcribed his diary and then gave a copy to the Yosemite Museum. Included with that copy were many of the expedition's photographs, but not the one of the Devils Postpile. Hutchings' original diary was held by Cosie's son William Mills, Jr., and is also now in the museum collection. Although without photographs, it contains some sketches not included in the transcribed copy. Hutchings had sent a number of photographs to the Secretary of the Interior to support the addition of Mount Dana to the newly created Yosemite National Park, including the one of the Postpile. The wayward photograph was unearthed at the National Archives, and we now have it as the cornerstone of this tale.

Hutchings' diary entry for September 19, 1875, provides the following description of the Devils Postpile: "The first thing that we did this morning was to visit the basaltic cliff for the purpose of taking a view. The more this formation is examined the more interesting it becomes. In the centre the columns are vertical and as regular almost as though carefully cut and placed in shape by a stone mason. One had fallen out of place and was leaning forward from the stack (so to speak) probably about 15 or 18 feet at top. This was shaped thus (give measurements of sides of square). On the easterly side they were leaning about in this shape. An immense mass of broken columns formed the debris (see picture). Climbing to the top we found the ends of the columns smoothed and polished, and presented the appearance of a mosaic or tesselated floor. This smoothing was done of course by glacial action" (FIG. 2).

In the margin of the diary Hutchings gives "the measurements of the Basaltic Trap Column leaning out" as 7½, 17½, 15, 12¾, 6¾, and 4 inches.

As far as we know, these observations precede by four years the first published mention of the Devils Postpile. A brief description appeared in a Fresno newspaper in 1879 after the Postpile was seen by a crew surveying a route for a road from Fresno Flats (Oakhurst area) to the gold mines at Mammoth.

Hutchings' "leaning forward" basalt column can clearly be seen in James' photograph (FIG. 3). At first glance it would appear to be leaning against a tree. This is an illusion, however, as closer examination indicates that the tree in

FIGURE 2
Mosaic pattern of glacially smoothed column tops. *Jim Snyder photo.*

FIGURE 3
Photograph of the Devils Postpile by W. E. James, 1875. Hand-written note on reverse of original photograph reads "Mountain of basaltic columns 50 ft long without a break, and only about 20 inches in diameter." *National Archives, Reston, VA.*

FIGURE 4
Detail of James' photograph of the Devils Postpile in the area of the leaning column. *National Archives, Reston, VA.*

question is actually growing along the top of the cliff tens of feet behind the column, as are the other trees in the photograph, and the column is freely leaning outward (FIG. 4). Also visible in the photograph are two figures. One, just to the right of the base of the leaning column, is probably Hutchings making his measurements on the column. A second figure at the top of the Postpile is probably Dr. Kellogg.

The precarious position of the leaning column invites physical analysis. From measurements on the photograph the column appears to be leaning outward at an angle of at least 20°. But how could a column of considerable length and weight be leaning out at such an angle? We speculate that the leaning column was buttressed at its base by some column stubs in front that were somewhat higher than the leaning column's basal fracture. In the photograph there appear to be quite a few such column stubs in the general vicinity.

The measurements provided by Hutchings of the leaning column's sides allow reconstruction of a probable cross-section of the column, which compares reasonably well with a rough sketch in Hutchings' original diary (FIG. 5). But the length of the column cannot readily be determined from the photograph. Consequently we organized an expedition of our own to obtain more data.

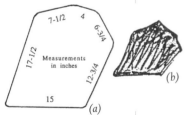

FIGURE 5
a: Cross-section of basalt column derived from Hutchings' measurements. Area is about 1.85 square feet. b: Sketch from Hutchings' diary. *NPS Yosemite Museum.*

James' photograph is sufficiently detailed to recognize individual columns visible today and we were able to locate the leaning column's former position quite closely. And at that point we discovered a column stub leaning outward from behind supporting stubs (FIG. 6). Was this what we were looking for? Because of access restrictions, we could not get to the stub itself, but looking down on it from the top of the Postpile it appeared to match the described cross-section (FIG. 5) quite well and was in the proper location. James' photograph strongly suggests that the leaning column originally reached to the top of the adjacent cliff. Lowering a tape measure to the top of the remaining stub gave a length of 28 feet. Estimating a length of 6 feet for that remaining stub gives a total length for the leaning column of about 34 feet.

If a 34-foot-long column were leaning out at an angle of 25°, its top would be about 15 feet away from the cliff face. If Hutchings' estimate of "15 or 18 feet" is accurate, then an angle of at least 25° would be required. James' photograph supports this conclusion. Figure 7 illustrates our reconstruction of the leaning column's setting.

The area of the column's cross-section is approximately 1.85 square feet. Assuming a uniform cross-section over a length of 34 feet, the volume of the column would be about 63 cubic feet. Assuming a density (specific gravity) for basalt of 3.0, the column's weight would be about 11,770 pounds, or roughly 6 tons.

FIGURE 6
Photograph of a column stub leaning forward from behind column stubs in front of it. *Photo by Gerhard Schumacher, USGS Photo Library.*

Given that this column was indeed leaning far outward in 1875 at the time of Hutchings' visit, how could it have arrived at its precarious position and remained there? Columns commonly become separated from the cliff face to lean slightly outward, where they become more and more susceptible to catastrophic failure. Such a failure occurred in 1980 when two slightly leaning columns collapsed during one of a series of earthquakes in the nearby Mammoth Lakes region. But those columns toppled catastrophically and wound up as broken fragments among the others on the talus pile. For a 6-ton basalt column to rest at an angle of 25° or more, one would surmise that it must have reached that position slowly, as any quick movement would likely cause it to snap off and join the talus pile immediately rather than later. This would seem to rule out an earthquake as a triggering mechanism—but would it really?

If the fractured base of the column was behind and below the tops of some slightly higher column stubs, the column might be jiggled slowly into its tilted position during a series of minor shakes until it stabilized against those stubs, postponing ultimate catastrophic failure of the column. Although the type and timing

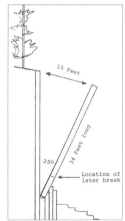

FIGURE 7
Schematic diagram illustrating probable setting of leaning column at time of Hutchings' visit, and the location of subsequent fracture resulting in collapse.

of a triggering event remain unknown, it is unlikely that the leaning column remained in its precarious position for very long. It was fortuitous that Hutchings happened by when he did, and had a photographer along.

Regardless of how this leaning column acquired its amazing position, it would be difficult to believe that such was even possible if it were not for the truly remarkable photograph taken by W. E. James while on J. M. Hutchings' 1875 expedition.

(First printed in *Yosemite*, v. 62, n. 1 [Winter 2000])

Tracking the Ice

From V to U—Glaciation and Valley Sculpture

BY N. KING HUBER

GLACIERS ARE POWERFUL AGENTS OF EROSION, CAPABLE OF GREATLY modifying the landscape. The Sierra Nevada has undergone multiple glaciations, although the exact number is unknown. Most of the glacial sculpture in the Sierra, however, was accomplished during the earliest and most extensive glaciation, known as the Sherwin glaciation, which ended a million years ago and may have lasted as long as 300,000 years.[1] At that time the granitic rocks that make up the bulk of the range were deeply weathered and offered little resistance to glacial excavation. Later glaciers, generally smaller, had to deal with the fresher and more resistant rock that was exposed by the action of Sherwin glaciers, and they were largely limited to "clean-up" activities, such as the removal of rock debris, or talus, that had tumbled from valley slopes during interglacial periods.

FIGURE 1
Glaciated Hetch Hetchy Valley with its broad, open floor. *Yosemite Research Library.*

Glaciers in mountainous terrain, such as the Sierra Nevada, largely follow and modify pre-existing stream valleys. Glaciated valleys tend to differ from normal stream valleys in two important ways. First and foremost, glaciated valleys tend to be straighter and less meandering than unglaciated stream valleys. Secondly, they tend to develop a

FIGURE 2
Unglaciated Merced River canyon below El Portal with its pronounced V-shaped cross-profile. *Photo by Dallas L. Peck, USGS Photo Library.*

U-shaped cross-profile rather than the V-shaped cross-profile characteristic of stream valleys beyond the reach of Sierran glaciers. Compare, for example, Hetch Hetchy Valley (FIG. 1) with the canyon of the Merced River immediately below El Portal (FIG. 2). Why this difference in form? Although existing pre-glacial rock structures play a part, the difference is largely due to the differing physical nature of the sculptors, water and ice.

River water, with its extreme fluidity, can flow fast enough so that its inertia, or resistance to change in direction, forces the water against the outside of river bends where it moves fastest. Thus rivers tend to erode the outside of their bends and to deposit sediment on the inside where the water flows more slowly. This behavior is beautifully illustrated by the meander pattern of the Merced River on the present Yosemite Valley floor. There the river has a low gradient, or slope, and is cutting into easily eroded alluvial materials only on the outside of river bends. Even in hard bedrock, a mountain river with high energy can cut deep canyons with entrenched meanders that remain sinuous, such as on the Merced River below El Portal.

In contrast, ice flows as a plastic solid. Glaciers move so slowly that inertial forces are negligible. A glacier will flow fastest, and erode fastest, where its surface slope is steepest, other things being equal. If a glacier occupies a sinuous valley cut by a river, the ice surface tends to drop more steeply on the inside of bends than on their outer sides (FIG. 3). Thus the fastest flow of ice tends to be on the inside of bends. Hence the insides of bends in the bed of a glacier tend to erode rapidly, and over time the glacier will tend to remove the topographic spur, or ridge, that forms the inside of the bend. In the process of eliminating the

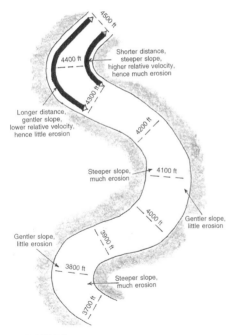

FIGURE 3
Diagram showing variation in steepness of ice-surface slope on opposite sides of bends in a valley glacier. Dashed lines are elevation contours on the upper surface of the glacier.

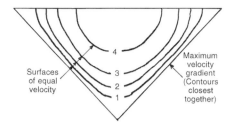

FIGURE 4
Schematic cross-section showing velocity contours within a glacier in a V-shaped valley. The maximum velocity gradient, where the shear stress is greatest, is part-way up the valley sides. After Johnson (1970).

original valley spurs, the glacier not only straightens the valley, but modifies the V-shape by broadening the valley floor.

Additional insight into the form of glacial valleys can be gained by examining the mechanics of glacial erosion. In considering the interaction of flow of a plastic material, such as ice, with the frictional effects of the walls of a relatively narrow V-shaped valley, it can be shown that the ice would have a maximum velocity gradient some distance up the sides of the valley walls (FIG. 4).[2] At this location the shear stress, a measure of erosive force, would be greatest. Thus the valley sides would be preferentially eroded there, and the V-shaped profile gradually changed to a U-shaped profile (FIG. 5).

Prominent planar fractures, known as joints, are common in granite and can exert significant control over glacial excavation, so not all glaciated valleys in Sierran granite will acquire a distinct U-shaped profile. If the rock has prominent joints, known as "sheet joints," sub-parallel to the sides of a V-shaped canyon, the glacier will simply pluck off the rock sheet-by-sheet, and so maintain a modified, but still V-shaped, canyon profile. Examples are the Grand Canyon of the Tuolumne River above Hetch Hetchy Valley (which contained the largest and most active valley glacier in the Sierra Nevada), the Merced Gorge between Yosemite Valley and El Portal, and Tenaya Canyon, all of which have retained roughly V-shaped cross-profiles. If joints trend across a canyon, or are otherwise irregular, glacial excavation will produce a more prominently U-shaped profile, such as in Yosemite Valley or Hetch Hetchy.

Once a glacier has sculpted a U-shaped bed, it can continue to excavate its bed to any depth. Where

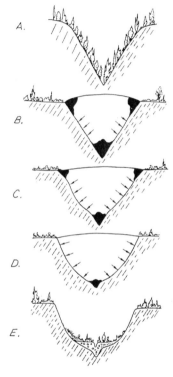

A. V-shaped mountain canyon.
B. V-shaped canyon visited by glacier.
C. Glacier erodes sides of canyon.
D. "Dead" regions disappear and entire side of canyon is rasped by rock-studded glacier.
E. Glacier disappears, leaving U-shaped valley.

FIGURE 5
Diagrams illustrating steps in the possible evolution of a U-shaped glacial valley. Black areas indicate "dead" areas of low velocity; arrows indicate places of intense glacial erosion. After Johnson (1970).

much ice has to flow through a narrower segment of the valley, or at the point of convergence of two glaciers, the ice must flow more rapidly through the slot to accommodate the combined volume of ice. Rapid flow tends to excavate deeply, and the glacier bed can be locally deepened to create a basin that later fills with water to form a lake after the ice melts. The deep bedrock basin in upper Yosemite Valley may have been carved by ice from the Merced and Tenaya glaciers that converged there during the extensive Sherwin glaciation. During the millennia following retreat of the Sherwin ice, that basin was mostly filled, in part by entrained material released by the melting ice, in part by stream-transported sediment, and in part by debris weathered and spalled from the valley walls.

The processes described above are best exemplified by the development of such long, straight, U-shaped valleys as those displayed by Stubblefield, Matterhorn, and Virginia Canyons in northern Yosemite National Park, and Lyell Canyon above Tuolumne Meadows (FIG. 6). These valleys are high enough in the range to have undergone multiple full-scale glaciations, including the last one, the Tioga glaciation, which peaked about 20,000 years ago.

FIGURE 6
The Lyell Fork of the Tuolumne River above Tuolumne Meadows. Note the gentle, sweeping curves of this broadly U-shaped glaciated valley. *Photograph by Robert W. Cameron. © Cameron and Company; used with permission.*

In many ways, Hetch Hetchy Valley, on the Tuolumne River some 15 miles north of Yosemite Valley, is a "fresher" example of a glacial valley than Yosemite. Even during the most recent Tioga-age glaciation, the Tuolumne River canyon was filled to its rim with ice. In that river's large drainage basin, three times the size of the one feeding ice to Yosemite Valley, many high-valley glaciers converged and completely filled Hetch Hetchy Valley, leaving lateral moraines thousands of feet above the valley floor.[3] Hetch Hetchy's walls are relatively clean, accentuating its classic U-shaped character. The valley has little talus because successive glaciers removed weak and weathered rock from the cliffs and rockfall debris from the valley floor.

Yosemite Valley's glacial history is quite different. The huge Sherwin glacier of one million years ago excavated the valley and overtopped its walls, but since that time no glaciers have completely filled the valley. The last and much smaller Tioga-age glacier scoured the lower parts of valley walls part-way down the valley and skimmed off talus and some valley fill, leaving the Bridalveil terminal moraine and El Capitan recessional moraine in its wake.[4]

As a result, over the last million years, the rock of the upper valley walls has weathered, joints have been enlarged, and rock has spalled off to form an irregularly sculptured surface, including the pinnacles and spires that we see today. With its angled, weathered walls and large talus accumulations, Yosemite Valley has lost much of the cleaner U-shape character that it might have once had. Because of post-glacial valley fill, its U-shape may actually have been visually enhanced. What Yosemite Valley has lost is the smooth, linear valley walls that are now sharply angled and deeply weathered.

John Muir glossed over these differences between Hetch Hetchy and Yosemite Valley. Calling all similar glaciated Sierran valleys, including Kings Canyon, "yosemites" as a generic term, he wrote that "Nature is not so poor as to possess only one of anything."[5]

At the same time, these differences puzzled Josiah Whitney who could see the evidence for an extensive glacier in Hetch Hetchy, but could not visualize effective glaciation in Yosemite Valley. He wrote: "The walls of the Yosemite on each side were carefully examined by the writer without his having been able to find on them any signs of smoothed, striated, or polished surfaces which could be unhesitatingly set down as the work of ice."[6]

Muir helped define and emphasize the erosive power of ice; Whitney was more receptive to the erosive power of water. Both were grappling with the problem of the origins of U- and V-shaped valleys. Since their time we have learned that mountains are sculptured not just by water, not just by ice, but over time by both in combination with other complex forces and conditions so striking that their results have been set aside as Yosemite National Park.

(First printed in *Yosemite*, v. 65, n. 4 [Fall 2003])

ACKNOWLEDGMENTS

This article draws on material compiled by my colleague, the late Clyde Wahrhaftig, for his geology classes at the University of California, Berkeley.

NOTES

1. G. I. Smith, V. J. Barczak, G. F. Moulton, and J. C. Liddicoat, 1983, *Core KM-3, a Surface-to-Bedrock Record of Late Cenozoic Sedimentation in Searles Valley, California*, U. S. Geological Survey Professional Paper 1256, p. 22.
2. For a mathematical derivation of this measure of erosive force see: Arvid M. Johnson, 1970, *Physical Processes in Geology* (San Francisco, CA, Freeman, Cooper & Company).
3. T. R. Alpha, Clyde Wahrhaftig, and N. K. Huber, 1987, *Oblique Map Showing Maximum Extent of 20,000-Year-Old (Tioga) Glaciers, Yosemite National Park, Central Sierra Nevada, California*, U. S. Geological Survey Miscellaneous Investigations Series Map I-1885.
4. N. K. Huber and J. B. Snyder, 2002, "A History of the El Capitan Moraine," Yosemite Association, *Yosemite*, v. 64, no. 1, p. 2–6.
5. John Muir, 1874, "Studies in the Sierra—Origin of Yosemite Valleys," *Overland Monthly*, June 1874, p. 496.
6. J. D. Whitney, 1880, *The Climatic Changes of Later Geological Times: A Discussion Based on Observations Made in the Cordilleras of North America* (Cambridge, MA, University Press).

A Tale of Two Valleys

BY N. KING HUBER

YOSEMITE NATIONAL PARK IS HOME TO TWO EXCEPTIONAL VALLEYS, Yosemite and Hetch Hetchy (FIG. 1). Yosemite Valley is renowned for its spectacular waterfalls and bold granite icons such as Half Dome and El Capitan and is a magnet for visitors from around the world. Hetch Hetchy Valley, although less well known and now the site of a reservoir for San Francisco's water supply, is also

FIGURE 1
The "Two Valleys" of Yosemite National Park: Yosemite and Hetch Hetchy. Looking eastward toward the Sierra crest, Yosemite Valley is just right-of-center. It extends directly up from near base of figure, to the valley head at the base of Half Dome. Hetch Hetchy Valley extends from its reservoir (3rd one from lower-left corner) diagonally up to the right. Topographic features are diagram-matic and exaggerated in this stylized graphic by Heinrich Berann. NPS *poster, 1988.*

quite remarkable. Indeed, John Muir, emphasizing the similarities between the two valleys, wrote that "Nature is not so poor as to possess only one of anything."[1]

The first comparison of the two valleys was presented at a meeting of the California Academy of Sciences in the fall of 1867 by Josiah Whitney, State Geologist and Director of the Geological Survey of California. Hetch Hetchy was characterized as "almost an exact counterpart of the Yosemite," and Whitney introduced a report by Charles Hoffmann, a member of his staff, who had explored Hetch Hetchy the previous summer. Hoffmann noted that "The scenery resembles very much that of the Yosemite, although the bluffs are not as high, nor do they extend as far."

Hoffmann described one waterfall (Tueeulala Falls) as having a sheer drop of 1,000 feet, and a second one (Wapama Falls) as a series of cascades dropping 1,700 feet. He remarked that "A singular feature of this valley is the total absence of talus at the base of the bluffs, excepting at one place in front of the falls. Another remarkable rock [Kolana Rock], corresponding with Cathedral Rock in Yosemite Valley, stands on the south side of the valley; its height is 2,270 feet above the valley." These early observations[2] have relevance to the discussion of the two valleys presented here.

The fundamental similarities that caught Muir's eye were that both Hetch Hetchy and Yosemite are broad but steep-walled valleys incised into the surrounding uplands, and that both have relatively flat floors traversed by meandering streams. Both valleys occupy similar positions on the western slope of the Sierra Nevada, with Yosemite's floor at about 4,000-feet elevation and Hetch Hetchy's slightly lower. Nevertheless, as noted by Hoffmann, Hetch Hetchy's valley walls, while impressive, are not as high as Yosemite's for the full length of the valley.

Although Hetch Hetchy Valley is nearly 4,000 feet deep near its head, downstream near its lower end the sheer cliff near Wapama Falls rises only about 1,600 feet from the valley floor (now submerged by the reservoir) to the upland plateau on the north. Kolana Rock, across the Tuolumne River on the south side of the valley, however, stands more than 2,000 feet above the valley floor, a smaller version of Yosemite Valley's 2,700-foot Cathedral Rocks. Hetch Hetchy's valley floor narrows upstream where its cliffs give way to the steep slopes of the Grand Canyon of the Tuolumne River, whereas Yosemite Valley's floor remains broad to its head near Half Dome.

In addition to these noted differences, it is even more significant that the walls of the two valleys are very different in appearance. Hetch Hetchy's walls are comparatively smooth and regular, while Yosemite's are jagged and irregular, with many pinnacles, spires, and deep re-entrants. These differences are graphically displayed by comparison of topographic maps of the two valleys (FIG. 2).

Hetch Hetchy Valley has relatively smoothly-curved elevation contours for most of its length; the only major indentation is where Tiltill and Rancheria Creeks breach the northern wall to enter the valley. In contrast, Yosemite Valley's contours emphasize the countless indentations and numerous pinnacles and spires jutting from the main walls.

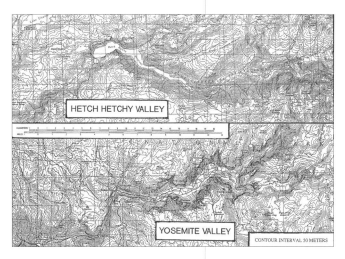

FIGURE 2
Topographic maps of the "Two Valleys." Note the comparatively smooth contours along the walls of Hetch Hetchy Valley as contrasted with the irregular, jagged ones in Yosemite Valley.[3]

Although we now know that both valleys owe their gross forms to glacial activity, Yosemite Valley's present morphology seems anomalous in that respect. The pinnacles and spires that punctuate its landscape, such as Lost Arrow, Sentinel Rock, and Cathedral Spires, could not have survived the erosive action of a glacier that filled the valley to the brim, as we know once occurred. How can we explain the presence of these striking features of Yosemite Valley, and thus the significant differences between the two valleys?

The answer to this question lies in the different glacial histories of the two valleys. Both histories had similar beginnings when the broad general shape of both valleys probably developed from glacial excavation during the Sherwin glaciation, a glacial epoch that ended nearly one million years ago. Sherwin-age glaciers filled each valley to its present rim, and locally beyond, with the Tuolumne glacier probably extending downstream a dozen miles below Hetch Hetchy to the Cherry Creek junction, and the Yosemite glacier as far as El Portal, some 10 miles downstream from Yosemite Valley proper.

The Sherwin was the most extensive, and evidently the longest-lived, glaciation documented in the Sierra Nevada. Later Sierran glaciations were of lesser areal extent and apparently briefer than the Sherwin, and here is where the glacial histories of the two valleys diverge.

FIGURE 3

Comparison of Tioga-age glaciers in Hetch Hetchy and Yosemite Valleys. In left diagram, Hetch Hetchy Valley lies beneath glacial ice from about the 6400-foot to 7200-foot ice-surface elevation-contours. In right diagram, glacial ice in Yosemite Valley reaches only as far as Bridalveil Meadow. Note that the ice tongue down Yosemite Creek (middle left) stops short of valley rim.[4]

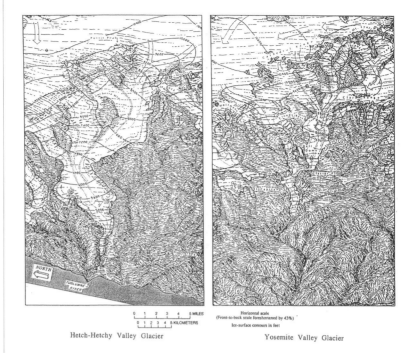

Hetch-Hetchy Valley Glacier   Yosemite Valley Glacier

Following the Sherwin, each major glaciation including the last one—the Tioga, which peaked about 20,000 years ago—produced glaciers that completely filled Hetch Hetchy Valley (FIG. 3). Moraines of Tioga age bounding Harden Lake, located on the south side of the canyon above the upper end of Hetch Hetchy Valley, indicate that the glacier was 3,700 feet thick there. Farther down, near the lower end of the Hetch Hetchy Valley, the glacier was 2,800 feet thick, with the north wall of the valley buried under ice delivered by tributary ice tongues flowing from the north down Rancheria, Tiltill and Falls Creeks to supplement the ice flowing down the main trunk of the Tuolumne River.

Thus, with each glaciation, including the latest Tioga, Hetch Hetchy's valley walls were being scraped clean and debris was being removed. Recall Hoffmann's early observation regarding the lack of significant talus in Hetch Hetchy Valley. There has been insufficient time since the Tioga for weathering and erosion to release significant amounts of talus from the "smoothed" valley walls or carve out pinnacles and spires from those walls.

In contrast, ice has probably not completely filled Yosemite Valley since the Sherwin glaciation about one million years ago (FIG. 3). The last glacier to enter Yosemite Valley, the Tioga, advanced only as far as Bridalveil Meadow. The extent of the somewhat earlier Tahoe glacier in the valley is uncertain, but evidence elsewhere in the Sierra suggests that it probably did not extend greatly beyond the Tioga. The fact that glaciers subsequent to the Sherwin failed to fill the valley to its rim has important consequences for the scenery.

From its terminus at Bridalveil Meadow, the ice surface of the Tioga glacier would have sloped upward toward the east end of Yosemite Valley, with the ice reaching a thickness of perhaps about 1,000 feet at Columbia Rock west of Yosemite Falls, 1,500 feet at Washington Column, and 2,000 feet in Tenaya Canyon below Basket Dome. Thus the Tioga and similar Tahoe glaciers could do very little to further modify or smooth the upper walls of Yosemite Valley proper. Above the ice surface of those glaciers, the valley walls have had a million years to be affected by the weather; joints have widened, rock has fractured and crumbled, and waterfalls and cascades have eroded alcoves and ravines. Thus, the pinnacles and spires that seem so anomalous for a glacial valley have had about a million years to form above the level of later glaciers, and so remain to amaze us today.

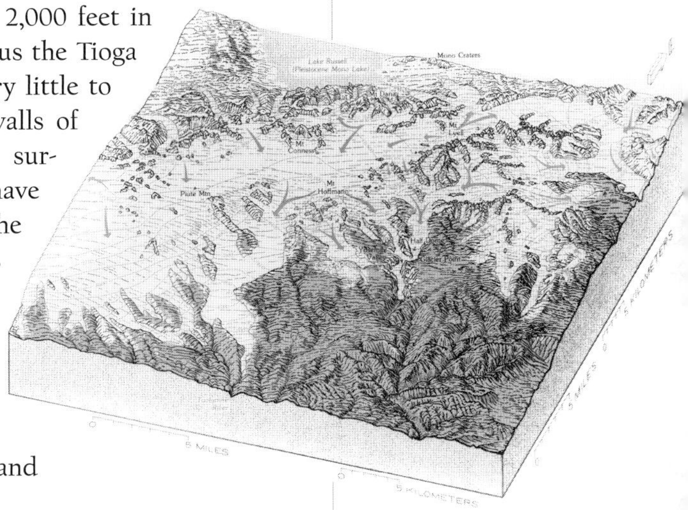

Meanwhile, back in Tenaya Canyon, the Tioga ice was closer to its source and thicker, rasping higher up on the valley walls and thereby smoothing them and removing irregularities so that no pinnacles or spires are found there.

Having ascribed the different geomorphic aspects of the two valleys to their different glacial histories, the next question is why those histories differ so. It was noted that the Tioga glaciation was much less extensive than the Sherwin glaciation that profoundly modified both valleys. The greater extent of the Tioga glacier in Hetch Hetchy, however, can be attributed to the fact that the drainage basin, or snowfall-catchment area, of the Tuolumne River system above Hetch Hetchy is more than three times as extensive as that of the Merced River above Yosemite Valley (FIG. 4).

As a result, the much larger icefield feeding the Tuolumne glacier was able to provide the volume of ice necessary to fill Hetch Hetchy Valley even though the Tioga glaciation was regionally less extensive than the Sherwin. This ice was delivered to Hetch

FIGURE 4
Extent of Tioga-age glaciers in Yosemite National Park. Arrows indicate direction of ice flow and show the much larger extent of the icefield feeding into Hetch Hetchy Valley. Note that the small glacier in Yosemite Creek (in center of figure) did not reach the rim of Yosemite Valley and thus did not contribute any additional ice to that valley.[5]

Hetchy Valley, both down the main trunk of the Tuolumne River, and by tributaries entering the valley from the north that were fed from the northeastern part of the Tuolumne icefield.

This tremendous influx of ice is what helped "clean out" Hetch Hetchy Valley. The smaller Merced River icefield was unable to provide sufficient ice to fill Yosemite Valley during the Tioga glaciation, even though supplemented by ice from the Tuolumne glacier that flowed southwest over several low passes in the Cathedral Range (FIG. 3) and over one from Tuolumne Meadows into Tenaya Canyon.

Having noted the significant differences between the two valleys, and having attempted to explain the why and wherefore of those differences, our tale cannot end without considering some of their consequences, especially with respect to Yosemite Valley itself. The Tioga-age glacier did little to further modify Yosemite Valley other than to remove fractured rock from the lower valley walls that had weathered and loosened since the previous glaciation. It also removed talus from the base of cliffs east of Bridalveil Meadow; all of the talus now there has accumulated in the last 15,000 years or so, after the Tioga glacier departed. For the past million years or so the rock walls of the valley that remained above the ice-level of the smaller post-Sherwin glaciers have weathered, joints have been enlarged, and rock has loosened and fallen to form the irregularly sculptured surface that we see today.

This geologic history provides the setting for frequent rockfalls. Every significant historical rockfall in Yosemite Valley has originated in vulnerable fractured rock derived from above the level scoured by the Tioga glacier. Some rockfalls have been quite large, but most are relatively small and gradually build up cones of debris below the more active sites. Thus the size of a debris cone can reflect the volume or the frequency of individual rock falls, or, most likely, a combination of both volume and frequency.

Less talus in Hetch Hetchy indicates less rockfall there, while in Yosemite Valley the opposite is true. The shattered rock high up on the east side of Middle Brother provides material for a debris cone at one of the most historically active rockfall sites in the valley. Both the 1996 "Happy Isles" and the 1998–9 "Curry Village" rockfalls added material to pre-existing debris cones that marked the sites of multiple, earlier events. Given the setting, such rockfalls will clearly play a major part in the dynamic processes that continue to shape Yosemite Valley.

(First printed in *Yosemite*, v. 66, n. 4 [Fall 2004])

NOTES

1. John Muir, 1874, "Studies in the Sierra—Origin of Yosemite Valleys," *Overland Monthly*, June 1874, p. 496.
2. C. F. Hoffmann, 1868, "Notes on Hetch-Hetchy Valley," *California Academy of Natural Sciences, Proceedings*, v. III, 1863–1867, p. 368–370.
3. Figure derived from: U. S. Geological Survey 1:100,000-scale *Topographic Map of Yosemite Valley, California*, 1976.
4. Figure derived from: T. R. Alpha, Clyde Wahrhaftig, and N. K. Huber, 1987, *Oblique Map Showing Maximum Extent of 20,000-year-old (Tioga) Glaciers, Yosemite National Park, Central Sierra Nevada, California*, U. S. Geological Survey Miscellaneous Investigations Series Map I-1885.
5. Figure derived from: N. K. Huber, 1987, *Geologic Story of Yosemite National Park*, U. S. Geological Survey Bulletin 1595 (reprinted by Yosemite Association, 1989), Figure 67.

General note: At the small scale of the maps shown here, it is not possible to clearly show all the place names mentioned. Other maps are readily available for those less familiar with Yosemite geography.

How Deep Is the Valley?

BY N. KING HUBER

MOST VISITORS, WHILE STROLLING ALONG THE RELATIVELY FLAT SURFACE of Yosemite Valley, have their eyes raised upward in awe of the valley's magnificent cliffs, towering pinnacles, and waterfalls, without thought of what might lie beneath their feet. But the surface they walk on, although altered by human activities, covers a considerable and varying thickness of sand, gravel, and boulders deposited by glaciers and streams over many millennia.

We are told that the gross shape of Yosemite Valley is the result of glacial deepening and widening of a section of the earlier stream-cut canyon of the Merced River. The present valley floor lies more than 3,000 feet below the upland surface on either side. But concealed beneath the materials filling the valley is a deep basin carved into the granitic bedrock. What is its shape and how deep beneath the present surface does it extend? This is the depth referred to in the title.

One of the first to seriously consider this question was François Matthes in his classic 1930 study, *Geologic History of the Yosemite Valley*. Matthes envisioned that the retreat of the last glacier from the valley left a lake, which he christened Lake Yosemite, that "occupied the entire length and breadth of the main Yosemite chamber. It was 5½ miles long, extending from the moraine dam at the [old] El Capitan Bridge to the great wall at the head of the valley . . ." "[I]t occupied an elongated basin scooped out in the rock floor of the valley by the ancient Yosemite Glacier."[1]

Matthes estimated a depth for that basin "ranging from 100 to not less than 300 ft," being deepest in the upper part of the valley opposite Yosemite Village. It is apparent that he believed that the last glacier, although smaller than earlier ones, scraped down to the bedrock floor of the original glacial excavation when he stated that "the glaciated rock floor of the valley (the bottom of ancient Lake Yosemite) represents excavating done since the beginning of the ice age."[2] The lake subsequently filled in with material left by the melting glacier, rock-waste shed from the cliffs, and sediment transported by the Merced River and Tenaya Creek.

Eliot Blackwelder, a glaciologist and contemporary of Matthes, believed that the fill might be much thicker because near-vertical glaciated slopes along the valley sides locally projected deep beneath the alluvium. At that time John Buwalda and Beno Gutenberg, of the California Institute of Technology, were experimenting with seismic-reflection equipment to determine subsurface structures. Blackwelder urged them to test his hypothesis by a geophysical survey of Yosemite Valley, which they did in 1935 and 1937.[3]

The seismic-reflection method depends upon the fact that a shock wave produced by a small explosion in the ground is partially reflected back from buried surfaces, such as the top of granitic bedrock. The time interval between the instant of the explosion and the return of the reflected wave can be accurately measured, and that time interval can be used to calculate the approximate depth to bedrock through the use of various assumptions and of observations from the seismic-reflection records.

The results of their survey were astounding. These results were first announced by Gutenberg and Buwalda in an abstract for a meeting of the Geological Society of America in 1937. They reported that the maximum depth to granite bedrock between the Ahwahnee Hotel and Camp Curry (Curry Village) was between 1,500 and 2,000 feet![4] This was followed in 1941 with a more descriptive article by Buwalda in *Yosemite Nature Notes*.[5] A complete technical report presenting all of the data and interpretations, however, was not published until 1957, with glaciologist Robert Sharp joining in to provide an "Introduction" and "Geological Interpretation."[6]

The results of the geophysical survey can best be abstracted from the 1957 published report, as follows: The bedrock floor beneath the surface of Yosemite Valley is an undulating surface with three separate basins. It slopes steeply from the head of the valley to its deepest point, about 2,000 feet deep, which lies between the Ahwahnee Hotel and Camp Curry (FIG. 1). Down-valley it rises

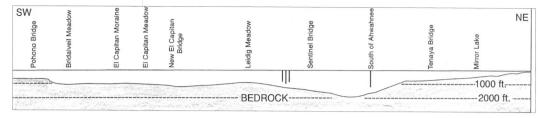

FIGURE 1

Longitudinal profile of bedrock basin beneath Yosemite Valley, measured along valley axis. The valley fill is estimated to be about 2,000 ft. deep in the basin south of the Ahwahnee. Dashed lines indicate depth below valley surface. Vertical lines indicate location and depth of wells described in text. The left cluster of wells is near Yosemite Lodge. The single well at right is in Upper River Campground. Bedrock profile after Gutenberg et al. (1957).

rapidly about 1,000 feet across a broad sill near Leidig Meadow. The second basin, about 1,375 feet deep, lies east of El Capitan Meadow, east of the El Capitan Bridge opposite Cathedral Spires. From here the floor rises gradually down-valley to perhaps 570 feet deep or less opposite Artist Creek, about ½ mile below Pohono Bridge. It might rise another 325 feet before the drop into a small basin more than 325 feet deep farther west at the Cascades.[7]

This complex configuration is shown on a map-plot that contours the thickness of the valley fill (FIG. 2). At the deepest location, the seismic data as interpreted by Gutenberg and Buwalda

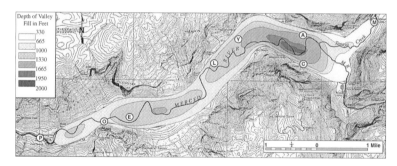

FIGURE 2

Map showing depth to bedrock in Yosemite Valley. Deepest area (below 1,950-ft depth) is located in center of valley between the Ahwahnee Hotel (A) and Camp Curry (C). Also note depression east of El Capitan Meadow (E) (below 1,330-ft depth). (L), Leidig Meadow; (M), Mirror Lake; (O), Old El Capitan Bridge at El Capitan moraine; (P), Pohono Bridge; (Y), Yosemite Lodge. The base map and river geometry are from the topographic map of Yosemite Valley by François Matthes (1907; Edition of 1934, with partial revision), available at the time of the seismic-survey compilation. Illustration modified from Gutenberg et al. (1957).

indicate that the bedrock basin has a broad U-shape concealed beneath the flat valley floor (FIG. 3). Adding this reconstructed trough beneath the present valley floor increases the overall depth of glacial excavation below the elevation of Glacier Point by more than 50 percent. Since the original glacial excavation about a million years ago, the upper valley walls have weathered, joints have widened, and rock slabs have spalled off so that those walls have become very irregular and no longer retain their glaciated geometry.

In 1956, William Colby, who had an advance copy of the technical report, wrote an article for *Yosemite Nature Notes* presenting a layman-oriented account that was more detailed than previously available. Of particular importance is a statement about additional seismic work in Yosemite Valley: "Many of the depth measurements obtained by Dr. Buwalda, assisted by Dr. Beno Gutenberg, were subsequently reshot from the same points . . . by a second geographical crew, using entirely different equipment. This group was under the supervision of Dr. Eliot Blackwelder. His results substantiate the earlier Buwalda figures."[8] The results of this later survey apparently were never formally published, but lend credence to the earlier work.

Through their association with the Sierra Club, Colby knew Matthes quite well back in the 1930s when the seismic work was being done. Colby recalled that when Matthes learned of the seismic results he realized that, if accepted, they would necessitate a profound modification of his expressed views on the amount of glacial excavation of the valley. Matthes then asserted that only actual borings would establish depth to bedrock to his satisfaction.[9] The seismic data actually reinforced Matthes' earlier concerns, expressed in 1913 after completing his topographic map of Yosemite Valley and at the beginning of his geologic studies, that "the existence of the rock basin [beneath Lake Yosemite] is purely inferential and is to be considered unproven until a series of borings along the whole length of the valley shall afford the necessary facts. It is to be hoped that some day such borings may

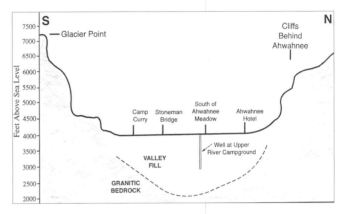

FIGURE 3
Cross section of Yosemite Valley between Camp Curry and the Ahwahnee Hotel. Dashed line indicates surface of granitic bedrock basin below the valley floor as interpreted from seismic-reflection data and plotted by Gutenberg et al. (1957). This reconstruction increases the overall depth of glacial excavation below the elevation of Glacier Point by more than 50 percent.

be undertaken; they would not merely serve to solve a problem of great local interest, but would contribute much-desired data regarding the still challenged eroding efficiency of glaciers."[10]

It was not until the 1970s and 1980s that four wells of significant depth were drilled in Yosemite Valley.[11] The most important is perhaps the one located in the Upper River Campground near the middle of the valley between the Ahwahnee Hotel and Camp Curry (FIGS. 1 & 3). This well failed to reach bedrock at a total depth of 1,200 feet, a depth consistent with the estimated bedrock depth of about 1,800 feet, indicated by the seismic-data contours. The three other wells were all located south and east of Yosemite Lodge near Yosemite Creek (FIG. 1). They reached depths of 870, 970, and 1,015 feet. The 870-ft. and 1,015-ft. wells did not reach bedrock. The 970-ft. well ended at either a "large boulder or solid granite," but as that well is only a short distance from the 1,015-ft. well, it is assumed that this well also did not reach bedrock. The seismic data suggest a bedrock depth of about 1,300 feet in this vicinity. While none of these wells actually establish a bedrock depth, they all indicate that the depths estimated from the seismic data are reasonable.

There is one more piece of evidence to consider. A regional compilation of gravity data depicts a gravity "low" between the Ahwahnee Hotel and Camp Curry, the same location as the seismically-determined maximum depth.[12] This "low" reflects the lower density of the valley-fill deposit compared to that of the underlying granite. While these gravity data have not been translated into specific depths to bedrock, they appear consistent with the well and the seismic data in indicating a great thickness of valley-fill at this locality. Thus, several lines of evidence indicate that Matthes vastly underestimated the extent of glacial excavation.

But we should not judge Matthes too harshly. The Lake Yosemite that existed following the demise of the last glacier in the valley was probably rather shallow, as that glacier probably excavated only a small proportion of the pre-existing sediment that had followed earlier glaciations. It would have taken an exceptional leap in imagination to propose the valley depth we now accept.

How can the "spoon-shaped" geometry of the deep basin be explained by glacial excavation? Robert Sharp, in his "Geological Interpretation" for the 1957 research paper, pointed out that a "likely factor was merging of the Tenaya and Merced glaciers, the two principal ice streams composing Yosemite Glacier. The location of the deepest point somewhat west of the junction of these two canyons is probably related to the greater thickness of ice there, at least 1700 m [5580 ft], and to the mode of flow within the glacier."[13]

A similar, added factor might be the narrowing at this point of the valley, which further forces the combined ice mass through a smaller opening. The influence of these factors is explained by recent mathematical modeling studies.[14] These studies indicate that "overdeepenings" result from an increase in ice discharge immediately below ice-stream junctions, which is accommodated primarily by increased ice thickness and hence increased sliding rate.

In recognizing that the last glacier ended within the valley, Matthes concluded that it "did not perform any significant share of excavating [Yosemite Valley]." And further, "The excavation of that basin clearly must have been accomplished by a much larger and more powerful glacier."[15] This early glaciation we now refer to as the Sherwin glaciation, which ended about 1 million years ago.[16]

Thus, while strolling along the valley floor and admiring the towering walls, try to imagine the valley when it was as much as 2,000 feet deeper!

NOTES

1. F. E. Matthes, 1930, *Geologic History of the Yosemite Valley*, U. S. Geological Survey Professional Paper 160, p. 103.
2. Matthes, 1930, p. 85.
3. J. P. Buwalda, 1941, "Form and Depth of the Bedrock Trough of Yosemite Valley," Yosemite Natural History Association, *Yosemite Nature Notes*, v. 20, no. 10, p. 89–93.
4. Beno Gutenberg and J. P. Buwalda, 1938, "Geophysical Investigations of Yosemite Valley" (Abs.), *Geological Society of America Proceedings for 1937*, p. 240.
5. Buwalda, 1941.
6. Beno Gutenberg, J. P. Buwalda, and R. P. Sharp, 1957, "Seismic Explorations on the Floor of Yosemite Valley, California," *Geological Society of America Bulletin*, v. 67, p. 1051–1078.
7. Gutenberg, Buwalda, and Sharp, 1957, p. 1051.
8. W. E. Colby, 1956, "The Latest Evidence Bearing on the Creation of Yosemite Valley," Yosemite Natural History Association, *Yosemite Nature Notes*, v. 35, no. 1, p. 3–8.
9. Colby, 1956, p. 5.
10. F. E. Matthes, 1913, "El Capitan Moraine and Ancient Lake Yosemite," *Sierra Club Bulletin*, v. 9, no. 1, p. 7–15; reprinted in *François Matthes and the Marks of Time* (San Francisco, Sierra Club), p. 55–62.

11. Yosemite Well Records, National Park Service, Water Resources Division, Fort Collins, CO 80525.
12. C. W. Roberts, R. C. Jachens, and H. W. Oliver, 1990, *Isostatic Residual Gravity Map of California and Offshore Southern California*, California Division of Mines and Geology, Geologic Data Map No. 7.
13. Gutenberg, Buwalda, and Sharp, 1957, p. 1074.
14. K. R. MacGregor, R. S. Anderson, S. P. Anderson, and E. D. Waddington, 2000, "Numerical Simulations of Glacial-Valley Longitudinal Profile Evolution," *Geological Society of America Bulletin*, v. 28, p. 1031.
15. Matthes, (1930), p. 74.
16. N. K. Huber, 2003, "Yosemite Falls—A New Perspective," Yosemite Association, *Yosemite*, v. 63, no. 1, p. 11.

After the Ice

Yosemite Falls— A New Perspective

BY N. KING HUBER

YOSEMITE FALLS WITH THEIR SPECTACUlar drop of 2,425 feet (including the Upper Fall, middle cascades, and Lower Fall) are world-renowned and an icon for Yosemite Valley (FIG. 1). They are truly unmatched and were recognized as such early on.

In 1851, the first descent into Yosemite Valley by non-Indians was made by the Mariposa Battalion under the command of Major James Savage. Lafayette Bunnell, who had served as a medical aide during the Mexican War and was called "Doc" by his colleagues, was one of the few literate members among the rough frontiersmen making up the group. He was greatly impressed by the valley, and around the evening campfire proposed the Indian name "Yo-sem-i-ty" for both the valley and the falls. Bunnell later recorded his impressions of the valley's scenic wonders in glowing terms. With respect to Yosemite Falls he noted that "comparison of the Yosemite Falls with those known in other parts of the world, will show that in elements of picturesque beauty, height, volume, color and majestic surroundings, the Yosemite has no rival upon earth."[1]

FIGURE 1
Upper Yosemite Fall now leaps from the hanging valley of Yosemite Creek. In the not-too-distant geologic past its water cascaded down through the prominent ravine immediately to the west (left). *Photo by N. King Huber, USGS Photo Library.*

FIGURE 2

This "Yo-hamite or Great Falls" drawing by Ayres was made into a lithographed poster by Hutchings in October, 1855, the first artistic representation of Yosemite Falls to reach the public. *Yosemite Research Library.*

James Mason Hutchings, an early promoter of Yosemite, organized the first party of sightseers to enter Yosemite Valley in 1855. Later that same year he published a lithograph based on a drawing of Yosemite Falls made at the time by Thomas Ayres, and thus introduced these spectacular cascades to the world for the first time (FIG. 2).

In his monumental study of Yosemite Valley, François Matthes wrote: "Surpassing all the other falls [of the valley] in height and splendor are the Yosemite Falls" and "they are easily [the valley's] most spectacular scenic feature. Even more than El Capitan and Half Dome they have given the Yosemite its wide renown."[2] But beyond their scenic impact, recognized by all, is the story of the falls' possible origin—one that is both fascinating and somewhat surprising.

Waterfalls cascading down a valley's walls from far above the valley floor have long been considered as evidence of a glacial origin for that valley. Indeed, little doubt exists that Yosemite Valley represents a profound glacially-driven modification of the pre-glacial Merced River canyon, because no other erosive agent could have accomplished such excavation. An ancient glacier that filled the valley to its present rim created the basic broad shape of the valley and gouged out a deep bedrock basin whose bottom locally lies more than 2,000 feet below the present valley floor.

The valley-forming glacial episode was named the "El Portal" glaciation by François Matthes in his Yosemite study because he estimated that the El Portal glacier advanced down the Merced River canyon to the vicinity of the community of El Portal, some 10 miles downstream from Yosemite Valley proper. Today, most geologists would correlate that glaciation with the "Sherwin" glaciation defined from studies along the east side of the Sierra Nevada. The Sherwin was the most extensive and longest-lived glaciation documented in the eastern Sierra. It may have lasted 300,000 years and ended about 1 million years ago.[3]

The enormous Sherwin-age glacier that shaped Yosemite Valley was fed from an icefield in the High Sierra, and was able to excavate the central chasm to a greater depth than smaller glaciers could erode in their side-entering tributary channels. When the ice left, some of the side valleys were left "hanging" with waterfalls at their confluence with the main valley. Since Sherwin time, most of the tributaries have eroded their channels back into the walls to leave little more than steep ravines with minor falls interrupted by chains of cascades, such as those at Sentinel Fall. Freefalling Bridalveil Fall is an exception, although it also has receded back into an alcove from its original position on the valley wall.

Later glaciations in the Sierra Nevada were of lesser area and apparently briefer than the Sherwin, but their actual number is uncertain. Matthes did recognize evidence for younger glacial activity, and he mapped the extent of what he called the "Wisconsin" glaciation, a name derived from the last glacial epoch in the northern mid-continent region of the United States. In the Sierra, his Wisconsin includes both the now-recognized "Tahoe" and "Tioga" glaciations, which probably peaked about 130,000 and 20,000 years ago, respectively. Much smaller than the Sherwin, glaciers of these later episodes did not come close to filling Yosemite Valley and thus did little to further modify or smooth its walls.

The sequence of post-Sherwin glaciations, however, probably contributed to the forming of the Upper Yosemite Fall that we know today. The Fall leaps from its hanging valley now, but as we will see, it has a more complex history than its neighbor falls, such as Bridalveil Fall.

Yosemite Creek is the largest stream flowing into the north side of Yosemite Valley and probably entered the pre-glacial Merced River canyon through a steep side ravine. Even after the Sherwin glacier excavated Yosemite Valley, Yosemite Creek continued to enter the main valley through that ravine, which lies just west of the site of the present Upper Fall (FIG. 1). At that time, the site of the present fall was fed by a very small drainage area between Yosemite and Indian Canyon Creeks and probably hosted only a minor ephemeral fall of short duration during spring runoff.

John Muir was impressed by the steep ravine, but did not recognize its pre-glacial existence when he noted that "there is a very deep cañon on the left of Yosemite falls . . . & I could not account for its formation in any other way than by supposing the existence of a glacier in the basin above."[4]

More importantly, Muir noted that Yosemite Creek is anomalous in that above the present Upper Fall it "is rather level . . .

FIGURE 3
John Muir noted that, unlike other large streams entering Yosemite Valley, Yosemite Creek flows across a rolling upland, descending gradually with a low stream gradient before cascading over the Upper Fall. *Photo by Frank Calkins, 1913, USGS Photo Library.*

and to the levelness . . . of this one, we in a great measure owe the present height of the Yosemite Falls. Yosemite Creek lives the most tranquil life of all the large streams that leap into the valley, the others, . . . while yet far from the valley, abound in loud falls and snowy cascades, but Yosemite Creek flows straight on through smooth meadows and hollows . . . biding its time . . . for the one anthem at the Yosemite . . ."[5]

It would appear that John Muir was the first to recognize that Yosemite Creek was unique in leaping out over the brink of Yosemite Valley (FIG. 3) rather than being ensconced in a "cañon," as were all of the other large streams entering the valley, including Bridalveil Creek.

In his Yosemite study, François Matthes concluded "that the broadly open notch west of the Upper Yosemite Fall was once the path of a stream flowing along the western margin of the [Yosemite Creek] glacier can hardly be doubted, for what appears to be an old stream channel leading to the notch is traceable along the west side of the [present] hanging valley for a quarter of a mile [FIG. 4], and 1,600 feet below the brink, at the foot of the incline on which the zigzag trail is built, there appears from beneath the débris a deeply cut stream channel, now dry, that joins the gorge of Yosemite Creek a short distance above the lower fall."[6] The trail leading north above the head of the notch continues to follow the old Yosemite Creek channel a little more than one-half mile to the junction with the present stream channel.

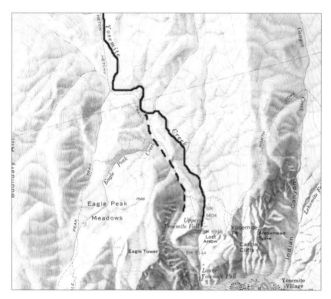

FIGURE 4
Yosemite Creek now and then. Today Yosemite Creek (heavy solid line) flows over the valley rim to create Upper Yosemite Fall. Before its postulated diversion, perhaps little more than 130,000 years ago, Yosemite Creek flowed down an older channel just to the west (heavy dashed line), from which it cascaded down through the steep ravine that is now the route of the Yosemite Falls Trail. *USGS.*

Much of the old stream channel in the steep ravine below the notch has been filled by rockfall since it was abandoned. However, the deep cut of the old channel mentioned by Matthes just above the Lower Fall is so pronounced that it shows clearly on the topographic map and was probably responsible for a small waterfall now present in the middle cascades. The site of this old channel, although largely obscured by vegetation, can be seen from below the Lower Fall (FIG. 5). It also can be seen from the Yosemite Falls Trail near where the trail turns upward into the actual ravine.

The presence of this lower ancient channel is critical to our interpretation of Yosemite Creek's history. It is lower in elevation than, or well below, the probable valley bottom of the pre-existing Merced River canyon prior to glacial deepening. It thus most likely continued to be the active channel for Yosemite Creek even after Yosemite Valley was glacially excavated a million years ago.

Matthes did not speculate on how or when Yosemite Creek was diverted from that old channel into its present channel to create Upper Yosemite Fall. He did, however, map a complex of glacial moraines, a series of arcuate ridges crossing the drainage above the valley rim (FIG. 6). These moraines consist of glacially-transported boulders and other debris that was deposited at the stationary front of the glacier occupying the valley of Yosemite Creek. As a glacial epoch comes to an end, and the rapidly melting front of the glacier retreats upvalley, episodic pauses in its retreat will result in the formation of a series of morainal ridges, termed recessional moraines, such as those mapped by Matthes. He attributed these moraines to his Wisconsin-age glacier that flowed down

FIGURE 5

View of Lower Yosemite Fall with site of ancient channel descending from upper left down through vegetated area. *Jim Snyder photo.*

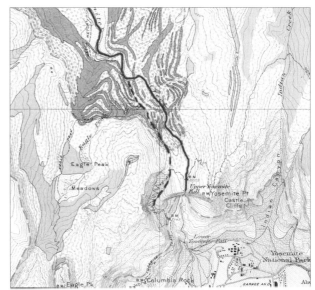

FIGURE 6

Glacial morainal complex of probable Tahoe-age on Yosemite Creek that caused stream diversion to present channel. Open circles indicate scattered morainal material. Darkly-stippled arcuate areas indicate complex of moraine ridges as mapped by François Matthes. *USGS.*

Yosemite Creek, but which apparently stopped about one-half mile short of the rim of Yosemite Valley itself.

While Matthes' moraines need to be remapped in light of our current knowledge of more numerous post-Sherwin episodes of glacial activity, they nonetheless offer a plausible explanation for the present-day location of Upper Yosemite Fall. The younger morainal deposits most likely blocked the old channel of Yosemite Creek, which was then forced to find a new path through the intricate complex of nested moraines and was diverted on to its present site at the lip of the cliff.

The notch of the present Upper Fall is much smaller than the notch of the old Yosemite Creek leading to the abandoned ravine because the site was originally fed by only a very small stream and because there has been much less time—maybe hundreds of thousands of years less—for erosion there since the diversion of the much larger Yosemite Creek to it. Still, the present channel is cut fairly deeply in massive bedrock and is certainly pre-Tioga in age. Thus the present Yosemite Creek channel is probably at least as old as Tahoe in age, possibly older if diversion can be attributed to an unrecognized older glaciation.

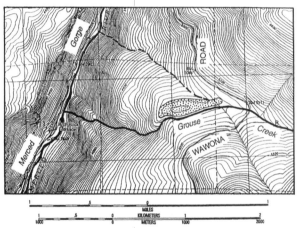

FIGURE 7
Grouse Creek, flowing from right to left (heavy solid line), skirts the uphill side of a glacial moraine deposit (stippled area). Prior to diversion by the blocking moraine, Grouse Creek followed the heavy, dashed line down to the Merced Gorge.

Lending credence to the above explanation are several specific examples of stream diversions by glacial moraines in Yosemite, both from the latest glacial episode and an earlier one. An example of the latter that was described by Matthes is on Grouse Creek just west of the Wawona Road (FIG. 7), where he observed: "Grouse Creek, it would appear, has been deflected from its original lower course by a heavy embankment of glacial débris—the left lateral moraine of the Yosemite [Valley] Glacier—and it now follows a new course, having broken through the embankment at a point half a mile farther south."[7]

Matthes did not locate this site on any of his published maps, but its probable location below the Wawona Road was deduced from a topographic map. My recent visit to the site with Jim Snyder, park historian, confirms Matthes' interpretation. The fortuitous preservation of this moraine is due solely to its diversion of Grouse Creek that has left it as a separate ridge along and outward from the canyon slope. This morainal deposit, some 2,000 feet above the Merced Gorge, is certainly old and is probably a product of the Sherwin glaciation. The degree of cavernous weathering of giant boulders mantling the deposit documents its considerable age (FIG. 8).

Within Yosemite Valley itself, prior to the final Tioga-age glaciation, the Merced River probably flowed westward near the center of the valley. Blocked by the cross-valley ridge of the terminal moraine constructed by the Tioga glacier at Bridalveil Meadow, the melt water broke through along the north margin of the glacier to form a spillway near the north valley wall where the river still flows today.[8]

Following upvalley retreat of the snout of the Tioga glacier to just west of El Capitan Meadow, the glacier stabilized and the El Capitan recessional moraine ridge of bouldery debris was constructed across the valley, again obstructing the drainage. This time the melt water broke through along the south side of the valley to form a spillway near the south valley wall where the river remains today.[8]

Matthes' Wisconsin glacial stage and his mapped morainal complex on Yosemite Creek may include both the Tahoe and Tioga glaciations. If the foregoing diversion scenario for Yosemite Creek is valid, Upper Yosemite Fall, with its "newly" hanging valley, is less than one million years old and may be little more than 130,000 years old. And what a spectacular addition to Yosemite Valley's architectural wonders it is!

(First published in *Yosemite*, v. 65, n. 1 [Winter 2003])

FIGURE 8
Giant, cavernously-weathered boulder mantling the Grouse Creek morainal deposit. Park Historian Jim Snyder provides scale. N. King Huber photo.

NOTES

1. Lafayette Houghton Bunnell, 1880, *Discovery of the Yosemite, and the Indian War of 1851 Which Led to That Event* (Reprint by Yosemite Association, 1990), p. 182.

2. François E. Matthes, 1930, *Geologic History of the Yosemite Valley*, U. S. Geological Survey Professional Paper 160, p. 18.
3. G. I. Smith, V. J. Barczak, G. F. Moulton, and J. C. Liddicoat, 1983, *Core KM-3, a Surface-to-Bedrock Record of Late Cenozoic Sedimentation in Searles Valley, California*, U. S. Geological Survey Professional Paper 1256, p. 23.
4. Robert Engberg and Donald Wesling, eds., 1999, *John Muir, to Yosemite and Beyond* (Salt Lake City, UT, University of Utah Press), p. 68.
5. John Muir, 1871, "Yosemite Glaciers," *New York Daily Tribune*, Dec. 5, 1871, p. 8, cols. 5–6.
6. Matthes, p. 112.
7 Matthes, p. 36.
8. N. King Huber and James B. Snyder, 2002, "A History of the El Capitan Moraine," Yosemite Association, *Yosemite*, v. 64, no. 1, p. 2–6.

A History of the El Capitan Moraine

BY N. KING HUBER AND JIM SNYDER

The El Capitan moraine is a prominent geomorphic feature in Yosemite Valley, but is little known to the average visitor as it is virtually hidden in a heavily wooded area west of El Capitan Meadow (FIG. 1). Although most visitors are unaware of the moraine, they wend their way past it on both the Southside and Northside Drives. The moraine consists of a low ridge nearly 1,500 feet long spanning the valley floor (FIG. 2). The only gap in the morainal ridge is at its south end near the valley wall where it has been breached by the Merced River. This gap is so narrow that it was chosen as the site for the original El Capitan Bridge constructed across the river in 1878.

The moraine, with its crest as much as 50 feet above the present stream bed, is composed of bouldery debris transported by a Tioga-age glacier and deposited at its front (FIG. 3). The Tioga glaciation, which peaked about 20,000 years ago, was the last major glaciation in the Sierra Nevada and the last to produce a glacier that entered Yosemite Valley. At its north end the moraine is less prominent as it is mantled by rock debris making up the alluvial fan constructed on the valley floor by Ribbon Creek.

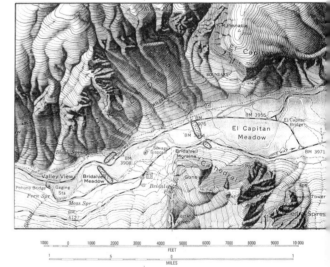

FIGURE 1

The El Capitan moraine is shown as an elongated, stippled area crossing the valley floor west of El Capitan Meadow. Also shown is a separate, arcuate moraine just east of Bridalveil Meadow. USGS.

The El Capitan moraine has sometimes been referred to as a "terminal" moraine, which would imply that it marks the Tioga glacier's furthermost advance down the valley. That honor, however, belongs to a moraine just east of Bridalveil Meadow, nearly a mile further down the valley (FIG. 4). The El Capitan instead is a "recessional" moraine constructed during a pause in the retreat of the glacier's ice front, or "snout," back up-valley from the Bridalveil Meadow site as the Tioga glaciation was approaching its end.

When the Tioga-age glacier disappeared from Yosemite Valley altogether, perhaps 15,000 years ago, it left behind a lake which François E. Matthes christened "Lake Yosemite."[1] It is likely that the advancing Tioga glacier, much smaller than the massive Sherwin glaciation a million

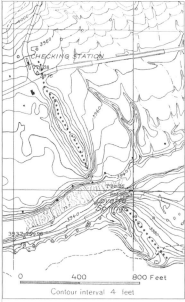

FIGURE 2
The El Capitan moraine is shown as an elongate ridge across the valley floor (crest marked by line of stars). Figure is from a 1:2,400-scale topographic map of Yosemite Valley Floor surveyed by the U. S. Geological Survey in 1919 and 1934. Note the presence at that time of a Checking Station adjacent to the moraine at the base of the Old Big Oak Flat Road. USGS.

FIGURE 3
Downslope side of El Capitan moraine with giant boulders exposed on surface. View is near Northside Drive where the moraine is relatively low. *Photo by N. King Huber, USGS Photo Library.*

years earlier, had excavated some of the pre-existing valley fill east of the El Capitan moraine to create a shallow lake basin. The lake was dammed by this moraine with the Merced River outlet flowing over a low spillway through the moraine near the south valley wall.

As the glacier retreated from the Valley, the Merced River and its tributaries, gorged with debris-laden meltwater from higher elevations in the mountain range, delivered large quantities of sediment to the lake basin. Deltas grew forward from each river and creek. With stream-borne sediment and rockfall debris from the weathered cliffs above the maximum extent of the Tioga glacier, the shallow lake was soon filled in, creating a relatively level valley floor.

The existence of glacial moraines on the floor of Yosemite Valley, and thus evidence for the past presence of a glacier in the valley, was first recognized in 1864 by Clarence King. King was a member of the Geological Survey of California, led by State Geologist Josiah D. Whitney, and King's observations were mentioned in Whitney's Geological Survey report published in 1865.[2] In a later publication, Whitney recanted and stated that "we have obtained no positive evidence that such [glaciation in Yosemite Valley] was the case. The statement to that effect in the "Geology of California," Vol. I., is an error, although it is certain that the masses of ice approached very near to the edge of the valley."[3] This denial of the presence and effects of glaciers in Yosemite Valley eventually precipitated a controversy between Whitney and John Muir regarding the origin of the valley.

In 1864 President Lincoln signed the act that granted Yosemite Valley and the Mariposa Grove of Sequoias to the State of California to "be held for public use, resort, and recreation; and shall be inalienable for all time. . . ." The following year, Frederick Law Olmsted, chairman of the state commission charged with overseeing the Yosemite Grant, presented a report to members of the Commission while they camped on the floor of Yosemite Valley.

This report, which proposed farsighted guidelines for management of the grant, included the statement that at "certain points the walls of rock are ploughed in polished horizontal furrows, at others moraines of boulders and pebbles are found; both evincing

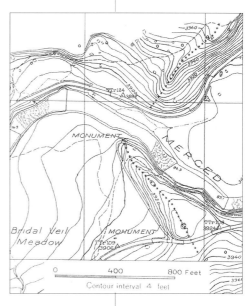

FIGURE 4

The Bridalveil Meadow moraine is shown as an arcuate ridge across the valley (crest marked by line of stars). *Map source as in Figure 2.*

the terrific force with which in past ages of the earth's history a glacier has moved down the chasm from among the adjoining peaks of the Sierras." This report was not released to the public, however, and remained generally unknown outside the Commission until its eventual publication in 1952 by Olmsted's biographer, Laura Wood Roper.[4] Whitney, although a member of the commission, was not at this Yosemite Valley gathering.

In the fall of 1864, Clarence King and James T. Gardner were hired by Olmsted to survey the boundary establishing the Yosemite Grant. In a later recounting of his observations made during this survey, King stated that "the markings upon the glacier cliff above Hutchings' house, had convinced me that a glacier no less than a thousand feet deep had flowed through the valley, occupying its entire bottom."[5] Presumably Olmsted had learned about Yosemite Valley glacial features from King. Thus the existence of the El Capitan moraine was known early on, despite Whitney's later denial of glaciers in Yosemite Valley.

Recognizing the moraine was one thing; understanding its effects and changes was another. Over the thousands of years since the last glacier left, and long before the Yosemite Grant was established, the Merced River began to cut its channel lower and lower through the moraine. This was accomplished by the river's transporting boulders, cobbles, and finer sediment downstream from the moraine, but boulders too large for the stream to transport, even in flood stage, were left behind to partially block the channel. The river lowered the stream level through the moraine and then cut through the lake-filling sediment too, forming a new, lower floodplain upstream from the moraine and leaving several remnants of the earlier floodplain high and dry.[6]

Even with this natural lowering, the moraine still affected human use of the valley, especially as tourists came in growing numbers. The moraine restrained spring runoff, closing roads and trails and inundating meadow pasture for domestic animals, the sources of both food and tourist transport. Wet meadows probably increased insect populations to uncomfortable levels each spring. Because the river furthermore was the valley sewer, and impounding of spring runoff retarded sewage removal, the moraine contributed to sanitation problems. During winter floods as well, the impounding effects of the moraine contributed to closed roads and trails, isolating hotels and cabins, and taking out bridges on a regular basis.

The river was not easy to cross before river levels went down. Until 1878 there were just two bridges across the river in Yosemite Valley, Sentinel Bridge and Folsom's Bridge below Rocky Point, both in the eastern part of the Valley. The commissioners decided to build a new bridge "to give a crossing further down, and enable tourists to make the circuit in carriages, as well as on horseback, without being obliged to ford the river." So the first El Capitan bridge was built in fall, 1878, on a location recommended by the Chief Engineer of the Southern Pacific Railroad "just above the rapids, and where there is considerably greater elevation to the banks than elsewhere."[7] In this way the stability and elevation of the moraine came to have a great effect on the circulation of people moving around the valley.

The winter of 1878–79 produced a good runoff, but one threatening some valley facilities. The commissioners, who supported channel clearing and reinforcement above Sentinel Bridge, also "had the rocks blasted above the rapids" at the new bridge on the El Capitan moraine "so as to lower the stream and relieve it at this point."[8]

Commissioner Galen Clark (FIG. 5), who was also the appointed Guardian for the Yosemite Grant, did the blasting and described it briefly in 1909: "[w]hen the El Capitan iron bridge was built in 1879 [sic] it was located across the narrow channel of the river between the two points of what remains of an old glacial terminal moraine. The river channel at this place was filled with large boulders, which greatly obstructed the free outflow of the flood waters in the spring, causing extensive overflow of the low meadow land above, and greatly interfering with travel, especially to Yosemite Falls and Mirror Lake. In order to remedy this matter the large boulders in the channel were blasted and the fragments leveled down so as to give a free outflow of the flood waters. This increased the force of the river current, which now commenced its greater eroding work on the river banks, and as the winding turns became more abrupt the destructive force annually increases. Some thorough system of protection should be promptly used to save the river banks from further damage."[9]

The blasting protected the new bridge at high water while also relieving upstream flooding. We do not know exactly what effects the moraine blasting had on such things. There was, of course, river bank erosion before 1879; in fact, the commissioners' work above Sentinel Bridge that same year was to prevent such

FIGURE 5
In the mid-1870s Galen Clark attended a Yosemite commissioners' meeting in San Francisco, where he had his picture taken by his old friend Carleton Watkins, in whose honor Mt. Watkins was named by Whitney's Geological Survey of California (Yosemite Research Library). Clark was also honored, with the designation of Mt. Clark.

erosion. But Clark noticed an increase in river bank erosion in the 30 years following the blasting and proposed additional protection of the banks from further erosion.

Evidence of Clark's "blasting" can be seen in the stream bed at low water. Many boulders with reasonably flat but irregular upper surfaces display radial fractures characteristic of those left by explosives used in blasting rock. A detailed examination of the stream bed was undertaken by James F. Milestone during low water in October, 1977. He states that an extensive reach of the stream bed, 263 feet in length, is strewn with dozens of blasted boulders. Milestone notes that while boulders in the stream channel range up to 12 feet in diameter (FIG. 6), the majority of those that appear to have been blasted range in size from four to five feet in diameter. He concluded that boulder removal may have lowered the stream channel at the moraine by three to five feet.[10] This conclusion is difficult to evaluate because it is based on a considerable amount of indirect evidence from upstream channel incision.

The exact result of Clark's blasting is impossible to determine after the fact. Any lowering of a significant reach of the stream bed through the moraine, however, would increase the stream's volume capacity for a given water surface elevation. But this would only be important during extreme flood stages. During the winter flood of January, 1997, El Capitan Meadow immediately upstream from the moraine was flooded, but the present channel and the overbank storage on the meadow were able to handle the stream flow without the crest of the moraine being overtopped. Examination along the length of the moraine crest yields no evidence, such as an abandoned channel, that it was ever overtopped other than at the Merced River gap since the moraine was originally formed. If so, it is unlikely that Clark's efforts made a significant difference in regard to the ability of the moraine gap to handle extreme flood volumes even greater than the 1997 event.

FIGURE 6
Smoothly flowing water in Merced River enters bouldery rapids on left, some 200 feet upstream from the center of the El Capitan moraine. Note size of large boulder, typical of many in the glacial deposit. This site is at the head of the rapids shown by stippled pattern on the river on Figure 2.
Photo by N. King Huber, USGS Photo Library.

Milestone also argued that Clark's blasting at the moraine lowered the ground-water table in the floodplain upstream. This is possible, because lowering the *average* elevation of the stream surface through the moraine would tend in the direction of increasing ground-water outflow as well as surface-water stream flow. Milestone presents evidence that river lowering produced corresponding side stream incision, also affecting water tables. In that way Clark did succeed to some extent in helping to drain meadows to an uncertain degree. Robert Gibbens and Harold Heady concluded, however, that "there is no evidence that the early spread of trees in meadows was facilitated by a lowering of the water table," pointing out that the invasion of trees had occurred in wet meadows a decade before 1879.[11] Natural channel and water table changes have been so complex, and flooding so variable, that it is difficult to quantify any specific results due to Clark's efforts. Nevertheless, following his study, Milestone suggested restoring the historic elevation of the El Capitan moraine-gap in an attempt to recreate the natural conditions existing before 1879.[12]

In 1992, the Water Resources Division of the National Park Service completed a study to determine the feasibility and potential effects of moraine restoration. After their analysis, the study team concluded that "the proposed moraine reconstruction will have only modest effects in channel capacities and associated ground water tables, and will be unlikely to reestablish historic (pre-blast) morphologic conditions."[13]

Stream systems are dynamic and ever changing, and the Merced River in Yosemite Valley is no exception. Ever since the El Capitan moraine was formed and the river flowed over the spillway near the south valley wall, the river has been gradually eroding its bed lower and lower through the moraine gap. Galen Clark's efforts, such as they were, only temporarily accelerated this ongoing process.

After the winter flood of 1950, a suggestion was made to further lower the El Capitan moraine gap to reduce future flood damage. The proposal ended in Washington, where NPS Chief Engineer Frank Kittredge thought that "it is neither practical nor desirable to think of changing the gradient or course of the Merced River in the Valley . . . and no one, I am sure, would ever consider the lowering of this natural dam or barrier as left by the glaciers." Kittredge referred to geologist John Buwalda's similar conclusion and stated that, in dealing with floods, "it is simply a matter of coping with nature on her own terms . . ."[14]

The present approach of the National Park Service rephrases Kittredge to "allow natural processes to prevail." Perhaps there will be no further fiddling with the El Capitan moraine.

(First printed in *Yosemite*, v. 64, n. 1 [Winter 2002])

NOTES

1. François E. Matthes, 1930, *Geologic History of the Yosemite Valley*, U. S. Geological Survey Professional Paper 160, p. 103.
2. Josiah D. Whitney, 1865, *Geology, Volume I*, Geological Survey of California, p. 421-423.
3. Josiah D. Whitney, 1868, *The Yosemite Book* (Geological Survey of California, NY, Julius Bien), p. 100; *The Yosemite Guide-Book* (Cambridge, MA, Harvard University Press, 1870), p. 112.
4. Frederick Law Olmsted, 1865, *The Yosemite Valley and the Mariposa Big Trees; A Preliminary Report* (Yosemite, CA, Yosemite Association, reprint 1995), p. 4.
5. Clarence King, "Around Yosemite Walls," Chapter 7 in *Mountaineering in the Sierra Nevada* (NY, James R. Osgood and Co., 1872), p. 152.
6. Matthes, p. 104; James F. Milestone, 1978, *The Influence of Modern Man on the Stream System of Yosemite Valley* (San Francisco State University, M.A. Thesis), p. 26–27.
7. Commissioners to Manage the Yosemite Valley and the Mariposa Big Tree Grove, *Biennial Report, 1878-1879* (Sacramento: State Printing Office, nd), p. 3–4.
8. Commissioners, *Biennial Report, 1878-1879*, p. 5.
9. Galen Clark, 1909, "Yosemite—Past and Present," *Sunset*, vol. 22, no. 4 (April, 1909), p. 396.
10. Milestone, 1978, chap. 5.
11. Robert P. Gibbens and Harold F. Heady, 1964, *The Influence of Modern Man on the Vegetation of Yosemite Valley*, California Agricultural Experiment Station Extension Service Manual 26 (Berkeley, University of California Division of Agricultural Sciences), p. 15-16.
12. Milestone, 1978, p. 158–166.
13. Gary M. Smillie, William L. Jackson, and Mike Martin, 1992, *Prediction of the Effects of Restoration of El Capitan Moraine, Yosemite National Park*, National Park Service Technical Report NPS/NRWRD/NRTR-92/10, p. 1.
14. Yosemite Archives, File 801-04, Part II, Yosemite National Park Storms, 1-1-51 to 4-30-53: NPS Chief Engineer Frank A. Kittredge to NPS Director, Washington, D. C., Jan. 2, 1951.

Exotic Boulders at Tioga Pass

BY N. KING HUBER

SOME YEARS AGO, TOGETHER WITH CARTOGRAPHER TAU RHO Alpha, Clyde Wahrhaftig and I published an oblique map showing the maximum extent of Tioga-age glaciers in Yosemite National Park (FIG. 1).[1] The Tioga, which peaked about 20,000 years ago, was the last major glaciation in the Sierra Nevada. Although each of us contributed to the map's construction, I was responsible for the arrows that show the direction of ice flow. In the Dana Meadows area, I added an arrow that split, with one arm directed west down the Dana Fork of the Tuolumne River and the other directed north over Tioga Pass into the Lee Vining Creek drainage basin. This is the double-pronged arrow in the upper center of the map (FIG. 1). Actually, the ice flow direction over most of the glaciated passes of the Sierran crest is indeed across the crest from the larger glaciers of the western slope and down the steeper canyons of the eastern

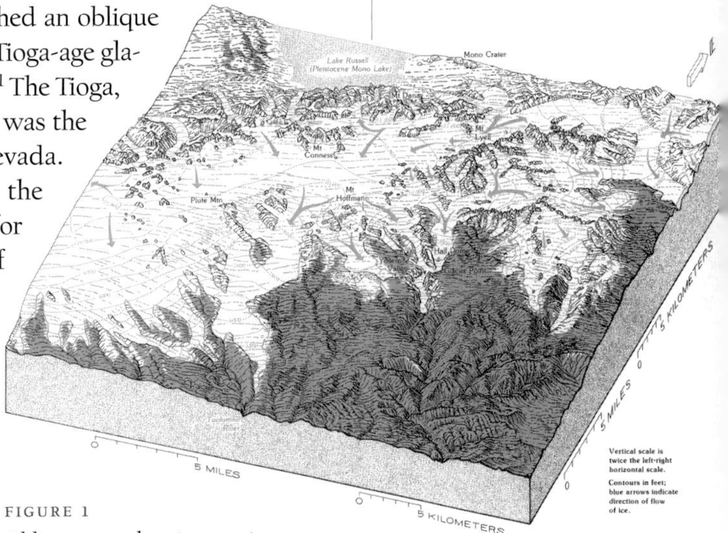

FIGURE 1
Oblique map showing maximum extent of ice during Tioga glaciation. Dana Meadows is located in the upper-central part of the figure where the split-arrow in question is shown. The double-tailed arrow west of Mount Hoffmann is heading down the Tuolumne drainage toward Hetch Hetchy. This version of the map appeared in Huber's Yosemite book[3] and was generalized from the original, larger-scale map by Alpha et al.,[1] which has much more detail.

slope. Unfortunately Clyde, who had determined the Tioga ice limits for our map, did not notice what I had done with this particular arrow and I was soon to regret it.

That an arm of the glacier flowed down the Dana Fork, as indicated by one arm of my arrow, is not in doubt. Clyde earlier had recognized that a boulder train of metamorphic rocks, derived from the Mt. Dana–Mt. Gibbs area and deposited along the Dana Fork, indicated that ice flowed westward down that drainage (FIG. 2). In addition, our colleague Malcolm Clark pointed out that parallel west-trending moraine ridges in the Dana Meadows area are medial moraines that extend back to the Dana-Gibbs area.[2] Their trend and position represent flow-lines of a glacier flowing westward down the Dana Fork.

Not long after the map was published, however, Yosemite naturalist Michael Ross told me that there were boulders of Cathedral Peak Granodiorite at Tioga Pass and asked me what they were doing there. This truly caught me by surprise and at first opportunity I went to check; sure enough, there they were. The Cathedral Peak Granodiorite (CPG) is a very distinctive variety of granitic rock containing large feldspar crystals as much as two inches in length scattered throughout a finer-grained rock matrix (FIG. 3).[3] On a weathered rock surface the more resistant feldspar crystals protrude above the more easily removed matrix (FIG. 4). Because of this distinctive texture the CPG is easily distinguished from other varieties of Yosemite's granitic rocks. The eastern boundary of the area occupied by the CPG lies well west of Tioga Pass (FIG. 2), and if an arm of the glacier had flowed down the Dana Fork there is no way that CPG boulders could arrive at Tioga Pass in the way indicated by the forked arrow that I had drawn. So where could those erratic boulders have come from?

Field examination of the Dana Meadows area showed that CPG boulders were most abundant in the immediate vicinity of Tioga Pass. These boulders decreased in abundance south and west through

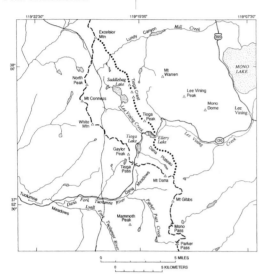

FIGURE 2

The Tioga Pass area and vicinity. The present Sierran drainage divide is shown by a heavy, dashed line. The heavy dotted line encloses the upper Lee Vining Creek drainage basin. The colored area on the west is underlain by Cathedral Peak Granodiorite.

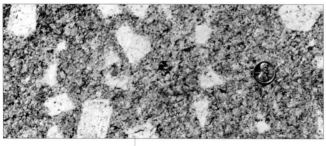

FIGURE 3

The Cathedral Peak Granodiorite. Its characteristic feldspar crystals are much larger than the other minerals in the rock matrix. Penny provides scale. *Photo by N. King Huber, USGS Photo Library.*

Dana Meadows, where the ice that carried them must have flowed southward through Tioga Pass and was compressed and turned westward by ice streams that issued from the Dana-Gibbs and Parker Pass Creek areas. The last and westernmost CPG boulder that I spotted was on the Dana Fork more than a mile upstream from the eastern CPG boundary (FIG. 2). The field evidence thus suggested that these exotic boulders were derived from somewhere north of Tioga Pass. But where?

Because the eastern boundary of the body of CPG continues north of the Dana Fork and crosses over the Sierra Nevada crest, a large area of CPG is exposed east of the Sierran divide in the vicinity of Mt. Conness (FIG. 2). Could Mt. Conness, which hosts one of the larger glaciers of the Sierra Nevada today, be a possible source for the boulders at Tioga Pass? Our reconstruction of the ice surface during the Tioga glaciation estimates that the valley glacier descending Lee Vining Creek from the Mt. Conness-Saddlebag Lake area was more than 1000 feet thick where that creek turns east below Tioga Lake to flow into Lee Vining canyon (FIG. 2). As this location is only about 500 feet lower in elevation than Tioga Pass, a 1000-foot-thick glacier could easily send an arm flowing south up and over that pass as well as an arm flowing east down Lee Vining Creek toward Mono Lake.

In summary, the erratic boulders of Cathedral Peak Granodiorite in the Tioga Pass area must have been derived from the Mt. Conness area. Such boulders would have been delivered from the Sierran crest area near Mt. Conness down to the west side of a valley glacier moving south along upper Lee Vining Creek. Their presence at Tioga Pass and in Dana Meadows indicates that during the Tioga glaciation ice flowed south up and over Tioga Pass as well as down Lee Vining Creek. This is the normal response of a glacier. It flows in the direction of the downhill slope of its upper surface, even if its base slopes uphill in that direction.

When I explained all of this to veteran Yosemite naturalist Carl Sharsmith he exclaimed, "I always wondered how those boulders got there!" He had recognized the puzzle long before anyone else and I was delighted to have been able to put the pieces together for him.

(First printed in *Yosemite*, v. 63, n. 2 [Spring 2001])

FIGURE 4
Weathered rock surface of Cathedral Peak Granodiorite with resistant feldspar crystals protruding above surface of more readily removed rock matrix. Penny provides scale. *Photo by N. King Huber, USGS Photo Library.*

NOTES

I would like to dedicate this article to the memory of Carl Sharsmith who told me about his excursions with François Matthes.

1. Tau Rho Alpha, Clyde Wahrhaftig, and N. King Huber, 1987, *Oblique Map Showing Maximum Extent of 20,000-Year-Old (Tioga) Glaciers, Yosemite National Park, Central Sierra Nevada, California*, U. S. Geological Survey Miscellaneous Investigations Series Map I-1885.
2. Malcolm M. Clark, 1976, "Evidence for Rapid Destruction of Latest Pleistocene Glaciers of the Sierra Nevada, California," *Geological Society of America, Abstracts with Programs*, v. 8, no. 3, p. 361.
3. N. King Huber, 1987, *The Geologic Story of Yosemite National Park*, U. S. Geological Survey Bulletin 1595 (Reprinted by Yosemite Association, 1989).

The Slide

BY N. KING HUBER, WILLIAM M. PHILLIPS, AND WILLIAM B. BULL

INTRODUCTION

"The Slide" on Slide Mountain in northern Yosemite National Park may well be one of the largest individual cataclysmic landslides in the central Sierra Nevada (FIG. 1), although very large debris flows are also known. The first known written description of The Slide is by First Lt. Nathaniel F. McClure, who came upon it in 1894 while exploring some of the canyons in Yosemite north of the Tuolumne River. He obviously was impressed by it: "After traveling three and one-half miles down the cañon, I came to the most wonderful natural object that I ever beheld. A vast granite cliff, two thousand feet in height, had literally tumbled from the bluff on the right-hand side of the stream with such force that it had not only made a mighty dam across the cañon, but many large stones had rolled up on the opposite

FIGURE 1

View looking southwest down Slide Canyon at The Slide on Slide Mountain near north boundary of Yosemite National Park. This giant rockslide, more than 840 feet wide, roared down with such energy that it climbed more than 120 vertical feet up the opposite side of the canyon. *Photograph by Robert W. Cameron, 1982. © Cameron and Company, used with permission.*

side. As it fell it had evidently broken into blocks, which were now seen of almost every size, piled one upon another in the wildest confusion. The smaller particles had settled between the crevices, leaving great holes among the larger blocks, some of which weighed many tons. To look at it, one might think that it had occurred but yesterday; but it was, in all probability, ages ago, as the ground just above the slide [upstream] is two hundred feet or more higher than that just below, showing that earth has accumulated on the upper side for many years."[1]

McClure's first impression was not surprising, as the landslide, or perhaps it is better referred to as a rockslide, may have occurred only about 150 years before he discovered it. The rockslide did pond the stream, but McClure overestimated the elevation drop across the dam; it is actually less than 40 feet. A small pond upstream from The Slide, also mentioned by McClure, has since been filled in to form a somewhat marshy meadow.

King Huber's interest in The Slide was aroused while studying aerial photographs during compilation of a geologic map of Yosemite National Park and the preparation of a book on the geology of the park, in which an oblique aerial photograph of The Slide (FIG. 1) was used as an example of a process of continuing landscape evolution.[2] In 1988 he finally was able to visit the site in the company of Jim Snyder, Yosemite Park Historian. They both were as impressed with the fresh appearance of this mass of jumbled rock as was Lt. McClure.

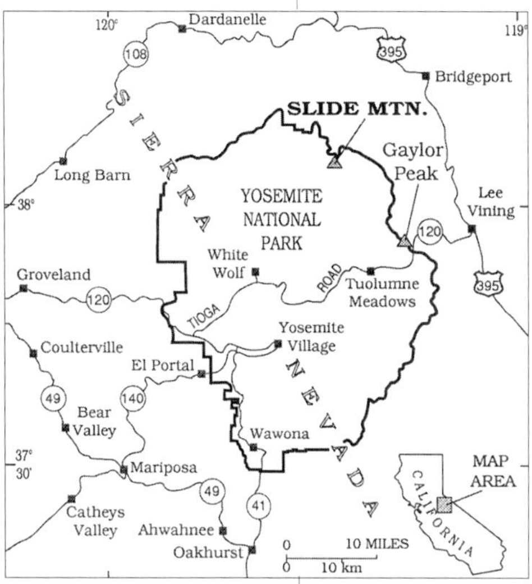

Two questions immediately came to mind: what might have been the triggering mechanism for the rockslide event, and when might it have occurred?

FIGURE 2
Index map showing location of Slide Mountain in northern Yosemite National Park. Near Slide Mountain, the park boundary follows the Sierran drainage divide.

PHYSICAL SETTING

The Slide occurs in a remote part of northern Yosemite National Park near the head of Slide Canyon, the glaciated valley of Piute Creek, only a little more than a mile from the Sierran drainage divide, which is also the northern boundary of the Park (FIG. 2). The elevation of the ridge on Slide Mountain at the head of the rockslide is about 10,600 feet, and Slide Canyon at the base of the rockslide is at about 9,200 feet, a drop of about 1,400 feet (FIG. 3).

At the time of the rockslide, blocks of granitic rock broke loose from the upper 480 feet of the west canyon wall over a width of about 840 feet and quickly became a rapidly moving mass of blocks streaming down the hill towards Piute Creek. This hurtling mass of granitic blocks behaved like a snow avalanche that starts out as a slab failure and then accelerates to dramatic speeds as it becomes fluidized. The rapid speed at which the rock avalanche moved is suggested by crude transverse waves and a raised rim where it stopped abruptly in a forest of hemlock trees. This rock avalanche achieved speeds estimated to be in excess of 140 miles per hour when it crossed Piute Creek to ramp up 120 feet in height on the opposite side of the valley (FIG. 4).

Talus blocks continue to accumulate on the upper slopes, but the morphology of the main rockslide indicates that it occurred as a single cataclysmic event. The granitic rocks on Slide Mountain are well jointed. The geometry of the joint system includes a series of lineaments that have been interpreted as the result of long-term lateral spreading toward Slide Canyon along steeply-dipping joints that parallel the slope face.[3] A Tioga-age (latest Pleistocene) glacier once filled Slide Canyon to the top of the headwall (FIG. 3). Following glacial retreat, more than 10,000 years ago, the

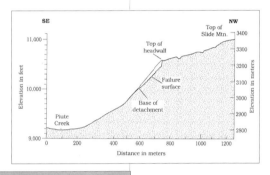

FIGURE 3
Schematic profile of the west side of Slide Canyon in the vicinity of The Slide. After Bronson and Watters, 1987.

FIGURE 4
View upstream toward The Slide in 1988 showing run-up of rock debris on east (right) side of Slide Canyon. *Photo by N. King Huber, USGS Photo Library.*

removal from the oversteepened valley slopes of the lateral support supplied by the ice would enhance such spreading. This type of gravitational deformation would have contributed to destabilization of the rock mass—a mass made ready for subsequent cataclysmic failure.

The volume of the rockslide has been estimated as at least 67 million cubic feet.[3] The deposit consists of an open framework of unsorted, angular, boulder-size blocks that average about 10 feet across, with many blocks in excess of 20 feet. The blocks have a freshly-broken appearance, and plant cover, except for small, scattered patches of lichens and mosses, is almost entirely lacking on the deposit.

SPECULATIONS ON ORIGIN

François Matthes concluded that most of the rock waste below cliffs in Yosemite Valley accumulated as talus derived by a continuous process of weathering and episodic spalling of rock debris from cliff faces. At the same time, he described "several masses of rock debris of enormous extent and wholly distinct from the ordinary sloping taluses . . . which can scarcely be accounted for save by the agency of earthquakes."[4] One such mass obstructed Tenaya Canyon to impound Mirror Lake. Matthes believed that this mass was derived from avalanches that fell from both sides of the canyon at the same time; if true, an unlikely event unless triggered by an earthquake. In their study of rockfalls in Yosemite Valley, Wieczorek and Jäger concluded that for events with documented triggering mechanisms, rain storms and rapid snowmelt triggered more numerous slope movements than earthquakes, but that earthquakes were responsible for a greater cumulative volume of material.[5]

With specific reference to the Owens Valley earthquake of March 1872, one of the strongest historic earthquakes to hit California, John Muir eloquently described a major rockfall in Yosemite Valley from north of Union Point above the Old Yosemite Village.[6] That earthquake also triggered a major rockfall from Liberty Cap; visitors to Nevada Fall can still see the light-colored scar up on the side of that dome that was the source area for the fall. The much smaller May, 1980, Mammoth Lakes earthquake sequence triggered several thousand rockfalls and slides throughout the central Sierra, including nine in Yosemite Valley.[5] Such evidence suggests that many of the very large rockfalls and slides in the Sierra are probably earthquake-generated.

CAN WE DATE THE SLIDE?

Three different approaches have been used in an attempt to determine the actual date of the rockslide—lichenometry, radiocarbon dating, and dendrochronology. Lichenometry involves the study of certain slow-growing and long-lived species of lichen and has been used to estimate the exposure age of rock surfaces. Attempted applications of lichenometry to The Slide only indicated a relatively "young" age.[7] Radiocarbon dating proved equally equivocal, with potential calendar-ages ranging from 1633 to 1802, as determined on a bulk wood sample from The Slide debris.[8] Indeed, lichenometry and radiocarbon are both statistical approaches and can only yield a general age, rather than a specific one. Dendrochronology, on the other hand, has the potential of providing a specific age, and thus is left as our best hope.

DENDROCHRONOLOGIC RESULTS

Dendrochronology, or tree-ring dating, is the study of the chronological sequence of annual growth rings with the goal of establishing the exact year in which each ring formed. Other things being equal, rings tend to be narrow in cold or dry years and wider in warm or rainy years, and over a long enough period of time, the sequence of narrow and wide rings is never repeated exactly. It is this recognizable sequence of wide and narrow rings that makes possible crossdating, or the matching of ring patterns in one specimen with corresponding ring patterns in another. A pattern established using a live tree (known year of outer ring) may be used to crossdate older dead trees or wood fragments that grew under similar environmental conditions.

With the hope of using dendrochronology to help date The Slide, during his visit King obtained two cross-section disks from tree stumps found among jumbled blocks of the rockslide on the valley floor. The trees were probably whitebark pine, a slow-growing alpine tree that presently grows near the upper part of the rockslide. These samples were compared with one from a live whitebark pine, but proved to be non-definitive and only suggested that the rockslide was older than the 1872 Owens Valley temblor.[9]

Subsequently a more concentrated effort to date The Slide was undertaken by Bill Phillips and Bill Bull from the University of Arizona. Their efforts proved to be more productive than King's. Samples from wood pinned beneath rockslide boulders were crossdated with a local tree-ring chronology that was constructed from living mountain hemlock trees growing in the vicinity of the deposit and which extends from A.D. 1488 to 1992. An impact scar on a living mountain hemlock tree on the margin of the deposit was also dated. Although the first-year scar-ring of that tree was not intersected by the coring device, the data restrict scar-formation to between 1735 and 1762. Two samples from logs pinned beneath boulders were especially informative in that they retained preserved bark, which assured us that the outermost annual ring at the time of the rockslide was preserved. Outermost rings of both samples date to 1739 and possess "latewood." This limits the time of slope failure to between the growing seasons of 1739 and 1740. An additional pinned log sample without preserved bark gives a minimum date of 1728 (maximum possible age).[10]

SUMMARY

Two questions were posed above: what might have been the triggering mechanism for the rockfall event, and when might it have occurred? It now appears that we have a reasonable answer for the latter question, which brings us back to the first one. Although a date of 1739–40 for the event rules out the 1872 Owens Valley earthquake, one cannot rule out an earthquake as the cause. It is still tempting to suggest seismic shaking as a trigger for the event because slope failures are commonly initiated by earthquakes and because The Slide is close to significant, frequently active faults along the eastern side of the Sierra Nevada. Because we cannot associate the slope failure with a known historic seismic event, its specific cause remains unknown. It could be seismic, climatic, or random failure of an unstable slope not tied to any one specific cause.

Nevertheless, along with Lt. McClure, we remain in amazement as we gaze at the awesome result of this catastrophic slope failure known as The Slide!

(First printed in *Yosemite*, v. 64, n. 3 [Summer 2002])

NOTES

1. N. F. McClure, 1895, "Explorations among the Cañons North of the Tuolumne River," *Sierra Club Bulletin*, v. 1, no. 5, p. 168–186.
2. N. K. Huber, 1987, *The Geologic Story of Yosemite National Park*, U. S. Geological Survey Bulletin 1595 (Reprint by Yosemite Association, 1989).
3. B. R. Bronson and R. J. Watters, 1987, "The Effects of Long Term Slope Deformations on the Stability of Granitic Rocks of the Sierra Nevada, California," *Engineering Geology and Soils Engineering Symposium, 23d, Logan, Utah, Proceedings*, p. 203–217.
4. F. E. Matthes, 1930, *Geologic History of the Yosemite Valley*, U. S. Geological Survey Professional Paper 160, p. 108.
5. G. F. Wieczorek and Stefan Jäger, 1996, "Triggering Mechanisms and Depositional Rates of Postglacial Slope-Movement Processes in the Yosemite Valley, California," *Geomorphology*, v. 15, p. 17–31.
6. John Muir, 1912, "Earthquake Storms," *The Yosemite* (Reprint by Sierra Club Books, San Francisco, 1988).
7. W. M. Phillips, W. B. Bull, and Thomas Moutoux, 1994, written communication, unpublished data.
8. R. J. Watters, 1992, written communication, unpublished data.
9. T. M. Yanosky, 1992, written communication, unpublished dendrochronologic data.
10. All data from Phillips, et al., 1994.

N. King Huber, Ph.D. Geologist Emeritus with the U. S. Geological Survey, mapped and studied the geology of the central Sierra Nevada for more than 40 years. His intimate knowledge of Yosemite's geology and geologic history made him an authority on the subjects, and as a geologic consultant to the National Park Service in Yosemite, he offered training to many park employees and visitors. In synthesizing the geologic story of Yosemite, Dr. Huber also drew on a century of geologic study by his U.S.G.S. colleagues and other geologists. He wrote *The Geologic Story of Yosemite National Park*, compiled the geologic map of Yosemite, and authored works on the geology of specific Sierra locations, such as Devils Postpile, as well as on Michigan's Isle Royale National Park.

The Yosemite Association is a 501(c)(3) nonprofit membership organization; since 1923, it has initiated and supported a variety of interpretive, educational, research, scientific, and environmental programs in Yosemite National Park, in cooperation with the National Park Service. Revenue generated by its publishing program, park visitor center bookstores, Yosemite Outdoor Adventures, membership dues, and donations enables it to provide services and direct financial support that promote park stewardship and enrich the visitor experience. To learn more about the association's activities and other publications, or for information about membership, please write to the Yosemite Association, P.O. Box 230, El Portal, CA, 95318, call (209) 379-2646, or visit www.yosemite.org.

Heyday Books, founded in 1974, works to deepen people's understanding and appreciation of the cultural, artistic, historic, and natural resources of California and the American West. It operates under a 501(c)(3) nonprofit educational organization (Heyday Institute) and, in addition to publishing books, sponsors a wide range of programs, outreach, and events. For more information about this or about becoming a Friend of Heyday, please visit our website at www.heydaybooks.com.